Conflict

Conflict, sadly, is part of our everyday life; experienced at home, in the workplace, on our TV screens. But is it an inevitable part of the fabric of our existence? In this volume, eight experts examine conflict at many levels, from the workings of genes to the evolution of galaxies.

Evolutionary biologist David Haig examines why we disagree with ourselves, and psychologist Simon Baron-Cohen asks whether differences between the average male and female mind must necessarily lead to misunderstanding. Anthropologist Richard Wrangham explores why chimpanzees and humans have evolved to kill, while archaeologist Barry Cunliffe describes the roots of warfare. Political scientist Lisa Anderson analyzes conflict in the Middle East, and broadcaster Kate Adie reflects on television reporting of war. The book concludes with industrial economist William Brown's discussion of conflict in labour relations, and an exploration of the creative and destructive effects of cosmic violence by physicist P. C. W. Davies.

THE DARWIN COLLEGE LECTURES

Conflict

Edited by *Martin Jones* and *A. C. Fabian*

CAMBRIDGE
UNIVERSITY PRESS

CAMBRIDGE UNIVERSITY PRESS
Cambridge, New York, Melbourne, Madrid, Cape Town, Singapore, São Paulo

Cambridge University Press
The Edinburgh Building, Cambridge CB2 2RU, UK

Published in the United States of America by Cambridge University Press, New York

www.cambridge.org
Information on this title: www.cambridge.org/9780521839600

First published 2006

Printed in the United Kingdom at the University Press, Cambridge

Typeface: Iridium LT Regular 10/14pt   *System:* QuarkXPress™   [SE]

A catalogue record for this book is available from the British Library

ISBN -13 978-0-521-83960-0 hardback
ISBN -10 0-521-83960-2 hardback

Contents

Figures

Tables

Acknowledgements

In addition to the named authors, a number of fellows and staff of Darwin College contributed to the lecture series and its publication. The editors would especially like to acknowledge the substantial contribution of Joyce Graham at every stage of the process, from conception of the series to preparation for publication, and convey their warm thanks.

Introduction: Arenas of conflict

MARTIN JONES AND ANDREW FABIAN

Equilibrium lies at the heart of many of our models of the way the world works. Arguments within a diverse range of disciplines proceed from the premise that the normal state of being is a harmonious one. But that premise is not secure. Disharmony and conflict may be far more than aberrations from a normal state; they may be at the heart of the system. The essays in this volume explore the theme of disharmony and conflict, and the alternative proposition that a variety of systems are fundamentally informed, not just by a tendency towards harmony and equilibrium, but also by conflict.

These essays were first delivered as the 20th Darwin College lecture series. They explore conflict at a range of levels and scales, from the workings of genes and chromosomes at one extreme, and the evolution of galaxies within a 'multiverse' at the other. Between these two extremes, they focus upon an intermediate scale, that of human society. In this introduction, we draw out some of the themes that have recurred in the course of these diverse contributions.

An evolutionary imperative

Conflict has a resonance with one of the central ideas of Darwinian evolution, the struggle for survival. This resonance is a recurrent theme of essays by Haig, Wrangham and Cunliffe, in their respective contributions on genomic conflict, the evolution of great apes, and the deep history of the human species. Wrangham and Cunliffe situate their arguments in the context of two contrasting images of humanity conveyed by figures of the eighteenth-century Enlightenment philosophy. For Jean-Jacques Rousseau, the natural state of humans was peaceful and benign, encapsulated in the idea of the 'noble savage'.

For Thomas Hobbes, humans and human societies were intrinsically violent. Cunliffe's 40,000-year survey of head wounds and weaponry steer him towards a Hobbesian conclusion, whereas Wrangham is less certain. In his study of human societies in the broader context of great apes, he sees strengths and weaknesses in both positions, and furthermore a variation in the placidity and violence of different primate species. Wrangham is persuaded that violence leading to death amongst humans and certain other great apes can be explained in adaptive terms, though there are important constraints on how it confers adaptive success. In great ape communities, lethal violence involves males killing males, and the killers operating in large enough groups to avoid putting their own lives at risk.

While interpersonal violence between males, in the case both of humans and chimps, may find some adaptive rationale, Haig draws our attention to a distinct arena of conflict. This arena lies not between individuals, but within them. This possibility is most simply exemplified by the case of one individual weighing up the costs and benefits of laying down his life to save three of his half-brothers. For the sake of some of his chromosomes and the genes they carry, he should certainly lay down his life. Other chromosomes within the same body are better served by keeping gallantry under control. Haig's essay explores the many ramifications of this seeming paradox. He considers various evolutionary scenarios in which different genes within the same individual may be favoured by different actions of the host organism. The distinct evolutionary strategies of maternally and paternally inherited genes routinely come into conflict within the individual who carries them, so for example they emphasize the development of different parts of the brain. This focus upon sexual difference brings us to a second theme, perhaps the most recurrent theme of the volume.

Conflict and men

Cunliffe argues that humans clearly have evolved to be gregarious, on the one hand, and aggressive and competitive, on the other. Indeed, it is the formation of communities, according to Wrangham, that provides the evolutionary rationale for lethal violence. Violent conflict between species is a long-recognized aspect of the Darwinian struggle for survival. It sometimes seems that such conflict within species is a very human attribute, but Wrangham draws our attention to the wide range of species that do attack and kill their own. He infers that killing a member of one's own species is a recurrent attribute of competition between rival communities of a single species. Following this argument, chimpanzees kill rivals whenever they can do so safely, because killing raises the likelihood of winning future battles. Among chimpanzees more than 90 per cent of adult victims of lethal inter-community aggression are male.

Many of the essays highlight the connection between violent conflict and men. In the background of this connection is Baron-Cohen's essay on the average male and female mind. While he stresses that men and women are widely dispersed in the manner they think and act, he observes a central tendency in each sex towards certain attributes of behaviour, attributes that can sometimes occur in extreme forms. From a wide range of observational studies of men and women ranging from maturity to a day old, and indeed some studies of foetuses, Baron-Cohen infers that the average male displays more 'direct' aggression, such as pushing, hitting and punching, than the average female, also noting that male-on-male homicide in contemporary society is thirty to forty times more frequent that female-on-female homicide. This is not to say that women are complete strangers to antipathy, but he argues that typical aggression between females takes more indirect forms, such as gossip, exclusion and cutting remarks. To quote Adie's essay on observing conflict, 'The history of war and insurrection, crime and domestic trouble, scenes of violence . . . is the story of man'.

There are certain recurrent rules of lethal conflict between males which apply in numerous species, including our own. An example is the rule of overwhelming advantage. Predatory killing proceeds from such a position of overwhelming advantage, in which the killing can be accomplished without placing the killers at significant risk. This is certainly true of chimp attacks, and of confrontations between a number of human societies. Within our own species, however, particularly in the context of stratified, hierarchical societies, that rule is transgressed in a different style of conflict, characterized by display, theatricality and performance.

Theatres of conflict

Adie recalls from first hand the launch of a cruise missile from a weapons platform off the coast of Serbia, in which the image of the launch was pasted up onto the front page of a New York newspaper before the missile had actually reached its target. This is probably the most extreme example of theatricality and performance among any of the contributions. That theme is explored by several essays, and a style of conflict that becomes distinct from its ancestral form. In exploring Middle Eastern conflict, Anderson emphasizes the dynamic between 'theatre' and 'audience', by observing that 'Israel is the battlefield but the war is in America. For Islamists like Bin Laden, America may be the battlefield but the war is in the Arabian peninsula.'

Within Europe, Cunliffe traces the roots of such theatricality of conflict back to the Bronze Age warriors of the third to second millennium BC. He sees emerging evidence for ceremonial display and simulated aggression, and a great diversity of weapons for hand-to-hand fighting and body armour. In early narratives drawing on this period the aggression of the opponents is contained by agreed rules and the contest is played out by the appointed champions in full view of all. Often, at the end of the day, everyone packed up and went home, but on some occasions passions would rise, the constraints would give and the rival fans would invade the pitch. Wrangham notes a number of unusual 'modern' traits of the soldier having a lot in common with the actor (detachment, illusion, etc.). The most unusual feature of all, in the context of comparisons with other species, is that the aggressor may risk his own life.

Wrangham connects this surprising development in conflictual strategy with a professional army and hierarchical communities in which orders can be given to override individual intuition. Both he and Cunliffe divide human communities into non-hierarchical and hierarchical forms, in which the former, including the Amazonian Yanamamö, the Maring and Dugum Dani from New Guinea, and the Andaman Islanders follow a more widespread pattern of intergroup violence, while hierarchical state societies, from the Bronze Age through to the present day, enact conflict in a theatrical manner, in which many of the key players are there as observers and in some cases controllers. Adie reflects on the various paradoxes of her central role in that collective observation, exploring the powerful impact on the process of observation on the structure of the performance itself.

An inevitable feature of theatres of conflict is the distinction between conflict on stage and off stage, between overt and covert action. A number of authors comment that the bloodiest and costliest conflict may be off stage. This applies locally: Adie distinguishes between the violence which for pragmatic and operational reasons remains unseen, and that which is suited to spectacle. It also applies globally, brought home by Anderson's comment that in the rank order of bloodiest twentieth-century conflicts, the principal theatre of global conflict, the Middle East, only appears at number 15. A recurrent theme of the essays is that the most lethal conflicts are by no means the most conspicuous. In the less lethal, but socially fundamental arena of labour conflict, Brown draws a parallel distinction between overt and covert conflict, again stressing the considerable suffering associated with the latter form, hidden from view.

Otherness and the 'nation'

Some manifestation of the 'other' is implicit in many forms of conflict, and those essays exploring contemporary societies explore a range of connections between conflict and the nation state, not all of which are self-evident. Anderson traces the birth of the inter-state system back to the 1648 Peace of Westphalia, its growth through the French Revolution, and what might be perceived as its Old Age in the breakup of the Ottoman Empire and the turbulent histories of the modern states of the Middle East. She explores the various conflicts between elective identities and national boundaries, single and divided loyalties, in a manner that would imply the world might be ready for an entirely new means of building communities. The often surprising power of national loyalties nonetheless recurs in the contemporary papers. Adie notes the considerable power of national allegiance in shaping and marketing wartime news. Brown alludes to the gulf that arose between the power of an international market, and unions that remained framed within national loyalties as a gulf that consumed the effectiveness of overt trade union action. National loyalties prevailed in spite of their economic disadvantage. To quote his contribution: 'The problem for trade unions is that, despite endless attempts, they are unable to organize collective action across national boundaries. Worker solidarity effectively stops at the frontier.'

Equilibrium, violence and life

We opened this introduction with the observation that equilibrium lies at the heart of many of our models of the way the world works, an observation that is challenged in a variety of ways in the following essays. Nowhere is that challenge more expansive than in Davies' consideration of dynamics within the universe. He highlights disequilibrium and infrequent events as key drivers at the largest scale of all, a dynamic that impacts upon phenomena at a range of smaller scales nested within, including the existence of life itself. He describes the take-home message of his contribution in the following way: 'violence has a creative as well as a destructive aspect, and that without exceedingly energetic and powerful processes that seem so awesome to human beings, life would be impossible'.

He has thus melded a portrayal of awesome devastation and violence with one of creativity and future life. In a volume that repeatedly revisits tension, suffering and destruction, there is much forward-looking and affirmation of the positive in these essays. A number conclude on such a positive note. Baron-Cohen argues that we should not treat the central tendencies of the male and female minds as constraints in individual choices and actions, and that knowledge of these differences can give us a much deeper understanding of routinely marginalized individuals. Adie highlights the growing sophistication of the global audience of conflict, implying a challenge to the parallel growth in sophistication of technologies of violence, assisted by rigorous observation. Brown emphasizes the continuing importance of democratic processes to challenge the iniquities of covert conflict. In other words, those authors immersed in the present and future of human societies hold out hope for enabled, sophisticated human agents to challenge and control the fundamental themes of conflict that pervade our world. It has to be said that the two authors looking back into the deeper human past conclude on more sanguine notes, with which we end this brief introduction.

Wrangham ends with the observation: 'If we take humans to be fundamentally similar to chimpanzees in their attitude to violence, we should be highly alert to the dangers inherent in major acquisitions of power. The biological intuition warns us that men will continue to look for opportunities to massacre their rivals, and that we should never drop our guard. The bad news is that we have to work to stop men from uniting to kill their rivals. The good news is that if we maintain balances of power, we can expect peace.'

Cunliffe concludes his essay with the words: 'If our all-too-brief look at the roots and development of warfare has taught us anything, it is how timeless and transglobal are the confrontations forced on us by the deeply embedded instincts that we aggressive humans have acquired through natural selection. Only a delicate tissue of social constraint, carefully nurtured, prevents our lives from becoming "nasty, brutish and short".'

1 Intrapersonal conflict

DAVID HAIG

Department of Organismic and Evolutionary Biology, Harvard University

In his *Principles of Psychology*, William James discussed five types of decisions. Most decisions he noted are decisions without effort, but in the

> *final type* of decision, the feeling that the evidence is all in, and that reason has balanced the books, may be either present or absent. But in either case we feel, in deciding, as if we ourselves by our own willful act inclined the beam: . . . If examined closely, its chief difference from the former cases appears to be that in those cases the mind at the moment of deciding on the triumphant alternative dropped the other one wholly or nearly out of sight, whereas here both alternatives are steadily held in view, and in the very act of murdering the vanquished possibility the chooser realizes how much in that instant he is making himself lose. It is deliberately driving a thorn into one's flesh; and the sense of *inward effort* with which the act is accompanied is an element which sets the fifth type of decision in strong contrast with the previous four varieties, and makes of it an altogether peculiar sort of mental phenomenon.
>
> (p. 1141)

After consideration of the kinds of decisions that are made with, and without, effort, James concluded that 'effort complicates volition . . . whenever a rarer and more ideal impulse is called upon to neutralize others of a more instinctive and habitual kind' (p. 1154).

Religious, literary and psychoanalytic texts abound with similar discussions of conflicts between our higher and lower natures, between passion and reason, between selfishness and concern for others, between immediate gratification and pursuit of long-term goals. We all are familiar with being caught on the horns of a dilemma – of wanting to make a phone-call, and simultaneously not wanting to make the call, of being torn between temptation and conscience – but evolutionary biology has had little to say on why our subjective experience should be organized in this manner. At first sight, the idea that we can be at war with ourselves appears paradoxical. If we are products of natural selection, superbly designed to maximize inclusive fitness, why do we often find it hard to make decisions and stick to them? A fitness-maximizing computer would simply calculate the expected utilities of the different alternatives and then choose the alternative with the highest score. Why should some kinds of decisions be more difficult to make than others? Is the subjective experience of effort merely a measure of the computational complexity of a problem, or is something else going on?

For William James, 'The existence of the effort as a phenomenal fact in our consciousness cannot of course be doubted or denied. Its significance, on the other hand, is a matter about which the gravest difference of opinion prevails. Questions as momentous as that of the very existence of spiritual causality, as vast as that of universal predestination or free-will, depend on its interpretation' (p. 1142). My aim in this essay is not to address such momentous questions, nor to shed light on really difficult questions such as how and why we have subjective experiences. Rather, it is to ask how one might begin to reconcile non-biologists' perception of the ubiquity of internal conflict with biologists' view of the mind as an adaptive product of natural selection. Internal conflict often *seems* maladaptive; consuming time, energy and repose. If so, why does it persist?

Three types of hypothesis could potentially resolve the conundrum of conflict within an adapted mind. First, one might argue that internal conflict arises from constraints on the perfection of adaptation; that evolved mechanisms work well on average but occasionally malfunction. We would be better off without internal conflict, but we are stuck with it. Second, one might argue that internal conflict is in some sense illusory; that the 'contending parties' have the same ultimate ends; and that natural selection has simply adopted an adversarial system as the best mechanism of arriving at useful truths. Finally, one might argue that internal conflict is 'real' and reflects a disagreement over ultimate ends between different agents that contribute to mental activity. I will reveal my hand at the outset. I believe that all three kinds of explanations, and their complex interactions, will play a role in an eventual understanding of internal conflict.

First, let me briefly consider non-adaptive interpretations. The precision of achievable adaptation is limited because natural selection is retrospective, fitting us to the past rather than the current environment; because the adaptive response to environmental change is limited by the pool of available variation and by time-lags until the origin of appropriate new mutations; and because selection is blind to very weak selective forces (i.e., chance plays a large role in who survives and reproduces when differences in adaptedness are slight). Some internal conflicts may merely reflect the imprecision of adaptation. An analogy can be made to the 'system conflicts' that occasionally cause my computer to crash: multiple functional programs are running simultaneously and occasionally make contradictory or ambiguous demands on the operating system; neither programmers nor natural selection have been able to eliminate all opportunities for malfunction. Our genomes evolve by a series of minor revisions to an old text just as the operating systems of computers evolve by the addition of new functionalities to old code. The analogy is, of course, limited. My computer does not, in fact, run multiple programs simultaneously. Instead, it is a serial machine that has only a single program running in its central processor at any particular moment, but switches rapidly among programs. Our brains, by contrast, are massively parallel processors with different subsystems simultaneously processing different kinds of data. Somehow this dispersed neural activity has to be integrated in coming to a decision. Perhaps 'conflict' could arise from imperfections in the process of integration.

Without doubt, our current environment presents us with novel challenges for which we lack specific adaptations. There were probably no opportunities in our evolutionary past to put aside resources for ten or twenty years, and then recover them with interest. Thus, retirement planning is a recent innovation for which we are unlikely to have evolved dedicated mechanisms. Instead, we employ general-purpose problem-solving machinery to make plans that come into conflict with more hard-wired responses. My rational resolution to save is thwarted by short-term impulses that fritter away income on ephemeral goods. (In this case, it is unclear that a comfortable retirement has anything to do with enhancing fitness. From a genetic perspective, our impulses may have it right.) Powerful narcotics are another novelty for which we are adaptively unprepared. An addict may strongly desire to be free of his compulsion, but may lack the will to override strongly maladaptive, albeit evolutionarily programmed, cravings.

Humans have evolved general-purpose problem-solving mechanisms, and the ability to learn from others, to compensate for the limitations of hard-wired adaptations. We are rational, cultural and instinctive beings. Sometimes these alternative sources of behavioural guidance promote different choices. Instinct summarizes the wisdom of past natural selection and recommends actions that have worked before under similar circumstances. Culture also summarizes wisdom from the past and can respond much faster than gene sequence to environmental change, but, from a gene's eye view, has the disadvantage of evolving by rules that need not promote genetic fitness. Reason can respond to unique features of the current situation and to weak selective forces, but may lack the historical judgement of either instinct or culture. Our passions, both positive and negative, are the carrots and sticks employed by genes to mould our actions to their ends. Reason may be a slave to the passions, but reason pursues pleasures as ends in themselves rather than as means to an end. (Coitus with rubber is an obvious example of reason circumventing genes' ends.)

Our ability to reason is an adaptive response to the imperfections of instinct, but this adaptive response must have its own imperfections, including inevitable clashes when instinct and reason offer conflicting advice. But if these clashes are unavoidable and recurrent, humans are expected to have evolved adaptive (but imperfect) mechanisms of resolving them. Questions of adaptation and of constraints on perfection are intertwined, and I already must abandon a rigid distinction between non-adaptive and adaptive hypotheses for the causes of internal conflicts.

How might a well-designed organism resolve conflicts between the dictates of instinct and the dictates of reason? Such an organism could have a limited ability to override instinct given strong enough reasons, with 'strong enough' calibrated to match the strength of past selection favouring the instinctive response. Very strong motivation would be needed for reason to prevail in decisions closely related to fitness, and for which instinct provides a powerful guide, but the threshold of motivation could be lowered when the prescripts of instinct are less strong. These considerations suggest an adaptive explanation for the feeling of effort in making certain kinds of decisions. Some decisions are hard to make because individuals in the past who made similar decisions with greater ease left fewer offspring. The strength of our will, no less than the strength of our muscles, can be shaped by natural selection. Moralists may derive some comfort from the muscular response to exercise.

Reality is, of course, more complicated than the above naïve model. Instincts are not unitary, nor is reason (itself a special kind of instinct). Different parts of the brain are specialized for different tasks and no part has access to the big picture. An arrangement in which mental modules compete for attention and influence could be a general organizing principle of the mind. Different modules process different kinds of data to produce recommendations for action. What a module communicates to the decision-making process may be no more than its preference among alternatives rather than a detailed justification of its preference. The preferences of different modules must then be aggregated to generate a choice. The economist Kenneth Arrow has shown that there is no method of aggregating preferences in situations of social choice that can be guaranteed never to violate basic axioms of rationality. His proof assumed that information was limited to the rank-ordering of preferences, and specifically did not allow interpersonal comparisons of the strength of preferences. Are there similar constraints on the perfectibility of intrapersonal aggregation of preferences?

In my subjective experience, choice is sometimes difficult because my internal rewards are not expressed in a common currency. Life would be simpler if one could compare expected degrees of orgasmic pleasure from giving in to a sexual temptation and from resisting it. Instead, the two courses of action have rewards that differ in kind.

There may be functional reasons for having more than one kind of reward – some rewards could be more suited to sustaining long-term projects, others for providing immediate gratification – but their existence renders intrapersonal comparisons more difficult. Multiple currencies would still allow facile comparisons among rewards if their exchange rates were well defined, but this does not seem to be the case. Why should this be so? If one part of my mind counsels one action and another part counsels the other, how is this disagreement to be resolved if there is no common unit for comparison?

Perhaps there are benefits, as well as costs, from not expressing all values on a single scale. Multiple currencies allow exchange rates to vary. This might be advantageous if the optimal weights to be attached to the pleadings of different internal voices vary from place to place, and time to time. A woman who yields to an extramarital passion may face very different consequences in New York and Riyadh. Multiple currencies could allow us to learn the appropriate exchange rates for our culture from our experience of which choices are rewarded and which punished during our behavioural maturation. If so, choices might be expected to become easier as we grow older and learn the norms of our culture (assuming these are not rapidly changing).

The discussion so far has concerned questions of adaptation and constraint: there may be adaptive reasons for allowing different internal voices to compete for attention but some expressions of this competition may be maladaptive. No mechanism is perfect. Pathological indecision may be just that: pathological. A model of the mind has been presented in which mental modules may express different preferences because they have different capabilities and process different inputs. In this model, the problem of preference aggregation reduces to a problem of determining the best weights to assign to the preferences of different modules. However, another possibility should also be considered. Agents with different interests can have different preferences even if they have access to the same information. If the self is an assemblage of agents with distinct interests, then internal conflict may reflect disagreements among agents over ultimate ends. What is adaptive for one, need not be adaptive for all. Agents may disagree about the relative weights that should be assigned to the recommendations of different modules. Thus, there need be no agreement on exchange rates, even when cultural norms have been fully internalized. In this view, decision-making would resemble the deliberations of a committee: sometimes consensus is achieved, sometimes one set of interests overrules the others, and sometimes the committee fails to decide.

I have in mind two sets of agents with stakes in the deliberations of the self. The first are genes. A gene's ultimate end is the propagation of its own copies. Put another way, the genes that we see today are those that have been successfully propagated in the past. Genes can be said to have *purposes* to the extent that they have properties that have promoted their own survival and replication. The second are ideas (or, to use Richard Dawkins' term, memes). Ideas may be propagated from other minds or generated afresh within a mind. (In fact, most of our ideas are hybrids that recombine content acquired from other minds with features generated in our own mind.) The ideas that reach our awareness have succeeded in competition with other ideas for attention. The ideas that colonize other minds have succeeded in competition with other ideas for expression by the transmitter and in competition with other ideas for perception by the receiver. Ideas can be said to have *purposes* to the extent that they have properties that have promoted their propagation from mind to mind. Such properties are the 'adaptations' of ideas.

Agents have conflicting interests whenever a choice that is best for the propagation of one differs from the choice that is best for the propagation of another. There has been conflict in my own mind over the writing of this essay. I have often been undecided about what to write and this indecision has preoccupied my mind, crowding out other concerns. Different ideas and forms of expression have competed for expression. Many sentences were entered into the manuscript, only to be erased and replaced by another sentence that I had temporarily discarded. Slowly a final text took form. What were the properties that made an idea successful in this competition? One of them was coherence with the rest of the manuscript. Another was my estimate as to how likely an idea was to attract your attention. I preferred pithy to dull formulations. Another criterion, I hope, was some degree of correspondence with reality, what one might call 'truth'. Now some of these ideas have entered your own mind, gentle reader. May they go forth and multiply.

Why do I care whether this essay is read? Why do people, in general, care about the propagation of their ideas? The fact that we care suggests there has been an evolutionary correlation between the transmission of ideas and of genes. That is, successful propagators of ideas have, on average, also been successful propagators of genes. Ideas can be useful, helping oneself and one's kin to survive in a challenging physical environment. And ideas can function in display: I want to impress you with how smart I am; I do not want to say anything silly; I want my ideas to spread so that I can bask in their reflected glory. Being a propagator of successful ideas must, on average, have translated into influence and control of resources. We have an instinct to generate and propagate ideas and this instinct creates the environment in which ideas can compete and evolve their own purposes.

'Good' ideas are persuasive and appeal to our genetically programmed biases, even if they do not promote genetic fitness. The correlation between genetic and memetic propagation cannot have been perfect and, at times, ideas may have propagated at the expense of genes, or genes at the expense of ideas. When a person willingly dies for a political or religious ideal, the urgings of ideas have triumphed over those of genes. But, when a charismatic preacher compromises his ministry for a brief sexual encounter, the entreaties of his genes have trumped those of ideas. Natural selection will favour the evolution of genetic biases against adopting ideas that reduce genetic fitness, but ideas evolve much faster than genes and will accommodate themselves to these new biases. Our choices are shaped by this interplay between the stubborn intransigence of genes and the supple agility of ideas.

But surely we should evolve a consistent set of genetic biases. Not necessarily, if we are also subject to conflict among genes. Biologists commonly assume that all of an individual's genes have the same interests because all have the same chance of being transmitted to that individual's offspring. For most choices, this is probably a reasonable assumption. However, there are subtle ways in which genes can have distinct interests and these can be the source of contradictory adaptations within the genome. Transposable genetic elements replicate at a faster rate than the rest of the genome. Nuclear genes are transmitted via eggs and sperm, whereas mitochondrial genes are transmitted only via eggs. If different genes have different rules of transmission, then an adaptation of one gene that promotes its own long-term propagation may not promote the transmission of the other genes with which it is temporarily associated. In the remainder of this essay, I will concentrate on the specific case of conflict between the nuclear genes we inherit from our fathers and those we inherit from our mothers. But, I will first need to digress on how natural selection acts on interactions among relatives.

We all die. A gene present in a neuron does not leave *direct* descendants. Only genes that are present in our sperm or eggs have any chance of producing copies that survive our death. Nevertheless, a gene in a neuron is a copy of a gene in the fertilized egg that gave rise to both our brains and our gonads and that has left its copies in both organs. Thus, genes in brains have evolved complex adaptations that promote the propagation of their *indirect* copies in gonads.

Our body contains multiple cell types, all with the same set of genes but with different roles to play in facilitating the transmission of gene copies via our gonads. However, if a gene in my liver can promote the transmission of its indirect copies in my gonads, then there is no reason why it cannot also express adaptations to promote the transmission of its indirect copies in the gonads of my relatives. When a mother provides milk to her baby, the genes in her breast promote the propagation of gene copies currently present in the baby's gonads not the mother's gonads. In this case, there is no guarantee that a gene in the mother's breast has indirect copies in a particular baby. Rather, a gene in her breast has only one chance in two of having an indirect copy transmitted to the baby. (The mother receives half her genes from her mother and half from her father. She transmits half to her baby, but this half is a mixture of the genes she received from her parents.)

Among cooperatively breeding meerkats (a kind of mongoose), daughters sometimes help their mothers by providing milk to their younger siblings. By this means, a gene in the daughter's breast may promote the propagation of its indirect copies via the mother's gonads, because the mother is relieved of some of the costs of lactation, and via the sibling's gonads, because the sibling is the direct beneficiary of extra milk. A gene in a daughter's breast has one chance in two of having indirect copies in the gonads of a sibling receiving extra milk, the same as the probability that the gene would have indirect copies in the gonads of one of the daughter's own offspring. If genes in a female's breast gain an equivalent benefit from providing milk to an offspring or to a sibling, why is sororal suckling not more common among mammals?

The equivalence of these two routes of promoting genetic fitness is predicated on the assumption that donor and recipient are full-sibs, sharing the same mother *and* father, as is usually the case in meerkats. Suppose, however, that the younger sibling has the same mother but different father. A gene that a daughter inherits from her mother has one chance in two of having an indirect copy in the younger sibling and one chance in two of having an indirect copy in the daughter's own offspring. The two routes of achieving fitness remain equivalent for genes of maternal origin. However, a gene that the daughter inherits from her father is absent from the younger sibling but has one chance in two of having an indirect copy transmitted to the daughter's own offspring. The two routes are not equivalent for genes of paternal origin. That is, maternal genes of the daughter are 'indifferent' as to whether an offspring or a maternal half-sib receives a benefit, but the daughter's paternal genes would 'prefer' her own offspring to receive a benefit instead of a maternal half-sib. Thus, under some circumstances maternal and paternal genes in a daughter's breast could 'disagree' over whether to supply milk to a half-sib.

The above simple example illustrates the general point that most relatives, excepting our direct descendants and full-sibs, are kin via our mother or via our father but not via both. Genes of paternal origin should 'care' about their effects on the father's side of the family, but not the mother's side whereas the situation is reversed for genes of maternal origin. Genes of maternal and paternal origin thus exist in different 'social environments' and may evolve different behaviours suited to their different circumstances. A biblical example can illustrate asymmetries of maternal and paternal kinship. Ishmael was Abraham's son by Hagar, an Egyptian slave-girl. While Ishmael dwelled with Abraham's extended family he would have been surrounded by many individuals with indirect copies of the genes he received from Abraham, but no individuals with indirect copies of the genes he received from Hagar, excepting Hagar herself. If Ishmael were to have taken some action that betrayed the household, but benefited himself, this action could well have had negative consequences for his genes of paternal origin but positive consequences for his genes of maternal origin. If Ishmael faced such a choice and similar situations had been repeated many times during human ancestry, one might expect his maternal genes to add a little more weight to the side of the scales favouring treachery but his paternal genes to add a little more weight to the side favouring filial loyalty.

Asymmetries of relatedness are greatest in an offspring's relations to its parents, and it is here that the potential for internal genetic conflict is predicted to be strongest. A daughter's maternal genes are definitely present in her mother, but the daughter's paternal genes are absent from her mother. For daughters' maternal genes, benefits to mothers are of equal value as benefits to self, but benefits to mothers are of little, if any, account for daughters' paternal genes. (Here, I gloss over some complexities due to variation in mating systems.) Significantly, although internal genetic conflict is present in a daughter with respect to her relations with her mother, the same is not true of the mother's relations with her daughter. The mother's maternal and paternal genes each have a 50 per cent probability of being present in the daughter. Thus, a child's feelings towards its parents are predicted to be more internally conflicted than the parent's feelings towards the child.

Consideration of conflict between maternal and paternal genes would be mere sophistry if not for the fact that some genes exhibit behaviour that varies depending on the sex of their most recent parental origin. This phenomenon is known as *genomic imprinting* because it is assumed that some mark, an imprint, becomes associated with a gene in a parent's gonad. This imprint must then be transmitted to the gene's direct copies in the next generation, identifying the gene as coming from a mother or father, but the imprint must be able to be erased in the gonads of offspring, so that the gene's copies can inherit the appropriate imprint in grand-offspring. An imprint is a contingent, rather than a fixed property of a gene, because a paternal gene in a daughter will be a maternal gene in the daughter's offspring. Typically, imprinted genes are actively expressed when inherited from one sex, but silent when inherited from the other. Thus, a past environment, whether a gene was in a male or female body in the previous generation, can influence the gene's expression in the current generation.

There is increasing evidence that imprinted genes can affect brain development and function in mammals. Barry Keverne and his collaborators at Cambridge have made an intriguing observation. The experimental details are beyond the scope of this essay, but these researchers were able to produce mice with two different kinds of cells in their bodies. Some cells were normal, with genes inherited from both a mother and a father, but other cells had genes inherited from only one sex of parent. Cells that lacked genes inherited from a mother were well represented in the hypothalamus of the mouse brain, but absent from the neocortex, whereas cells that lacked genes inherited from a father were well represented in the neocortex, but absent from the hypothalamus (Figure 1.1). These results suggest that maternally and paternally inherited genes perform different roles during normal development of the mouse brain. The results also hint that paternally derived genes of mice may favour relatively greater weight to be assigned to hypothalamic preferences in decision-making whereas maternally derived genes of mice may favour relatively greater weight to be assigned to neocortical preferences. (As a gross over-simplification, the hypothalamus controls 'visceral', and the neocortex 'cerebral', responses.)

There is much that I could say about why paternal genes of mice should favour hypothalamic development and maternal genes should favour neocortical development, but almost all I would say would be wrong. The phenomenon and the genes involved are still too poorly known to make anything other than wild speculations. For current purposes, I merely want you to entertain the possibility that situations

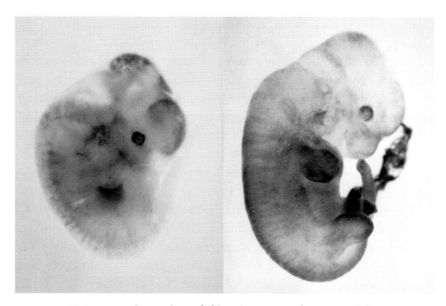

FIGURE 1.1: Comparison of chimeric mouse embryos containing gynogenetic and androgenetic cells. In the gynogenetic/normal chimera (left), the dark-staining gynogenetic cells are well represented in the developing brain but poorly represented in the small body. In the androgenetic/normal chimera (right), the dark-staining androgenetic cells are poorly represented in the developing brain, but well represented in the large body. (© Barry Keverne)

exist in which maternal and paternal genes within an individual have conflicting interests, so that I can ask how such conflicts might be expressed within the mind.

Imprinted genes would be expected to influence broad behavioural tendencies and personality traits, rather than micro-managing every individual decision. In the case of Ishmael contemplating treachery, Ishmael's genes cannot be expected to have any awareness of the details of his particular dilemma. Rather his genes would have instinctual information about outcomes of similar choices in the past when more or less weight had been given to the recommendations of temptation versus conscience. If, for example, maternal genes had benefited, on average, when relatively more weight was given to temptation, then maternal genes would be expected to evolve adaptations to promote a tendency to succumb to temptation. But these adaptations of maternal genes would be opposed by adaptations of paternal genes to increase the suasive power of conscience. This internal tension could be played out during development with the two sets of genes favouring different relative growth of different brain structures or it could be played out during brain function with one set of genes enhancing and the other damping the amplitude of particular kinds of neural signal.

What are the internal factors that determine our choices? Clearly, our natures are strongly influenced by the unique set of genes we inherit from our parents, but they are also influenced by the set of beliefs and memories that we have accumulated during our lives. Our choices are influenced by ideas just as much as by genes. Most choices are simple, but we sometimes need to choose in situations where genes and ideas provide no clear guidance or where different internal voices offer different advice. Conflicts can exist among ideas, among genes, and between genes and ideas. The Self can be viewed as the arbiter that mediates among the conflicting parties and then decides. We are free actors at least in the limited sense that no single set of interests exclusively determines our choices. We are also free in the sense that no one, not even ourselves, can predict with complete accuracy how we will choose in all situations.

FURTHER READING

K. J. Arrow, *Social Choice and Individual Values*, 2nd edn. New Haven: Yale University Press 1963.

R. Dawkins, *The Selfish Gene*. Oxford: Oxford University Press, 1976.

D. C. Dennett, *The Intentional Stance*. Cambridge, MA: MIT Press, 1987.

D. Haig, *Genomic Imprinting and Kinship*. New Brunswick, NJ: Rutgers University Press, 2002.

S. H. Heap, M. Hollis, B. Lyons, R. Sugden, and A. Weale, *The Theory of Choice*. Oxford: Blackwell, 1992.

W. James, *The Principles of Psychology*. Cambridge, MA: Harvard University Press, 1983. [This is a critical edition of the 1890 classic. The passages quoted in my essay contain some minor corrections from the text of the first edition.]

E. B. Keverne, R. Fundele, M. Narasimha, S. C. Barton and M. A. Surani, 'Genomic imprinting and the differential roles of parental genomes in brain development', *Developmental Brain Research* **92**, 1996, 91–100.

M. Scantlebury, A. F. Russell, G. M. McIlraith, J. R. Speakman and T. H. Clutton-Brock, 'The energetics of lactation in cooperatively breeding meerkats *Suricata suricatta*', *Proceedings of the Royal Society of London* **B 269**, 2002, 2147–53.

2 Sex differences in mind

SIMON BARON-COHEN

Department of Experimental Psychology and Psychiatry, Cambridge University

Why should the topic of sex differences in mind have any relevance to the theme of conflict, the focus of this book and lecture series? In my view, there are two links.

First, some consider that the male and female mind are so intrinsically different that conflict in some form – for example, mutual misunderstanding – is inevitable. I will argue that there are interesting differences between the *average* male and female mind but that recognizing these need not lead to conflict and instead could lead to mutual respect of difference. In using the word 'average', I am from the outset recognizing that such differences may have little to say about individuals, a point I will return to. In addition, the features that define the average male and female mind are not like chalk and cheese. The average male and female mind contains both chalk *and* cheese, as it were. The differences are subtle, and are to do with the relative proportions of chalk and cheese in the typical male and female mind. Fear not. I will get away from this metaphor quite soon, and define the qualities that we can quantify. But it does mean that because the male mind is not all chalk, and the female mind all cheese, the two sexes are not destined to be mystified by the other.

Second, the field of sex differences in psychology in the 1960s and 1970s, was so conflict-ridden as to make an open-minded debate about any possible role of biology contributing to psychological sex differences impossible. Those who explored the role of biology – even whilst acknowledging the importance of culture – found themselves accused of defending an essentialism that perpetuated inequalities between the sexes, and of oppression. Not a climate in which scientists can ask questions about mechanisms in nature. Today, the pendulum has settled sensibly in the middle of the nature–nurture debate, and scientists who care deeply about ending inequality and oppression can at the same time also talk freely about biological differences between the male and female brain and mind.

The conflict that characterized the 1960s and 1970s has not entirely evaporated, as witnessed by the global reactions in the media to Harvard President Professor Lawrence Summers' comments in 2005 when he was understood to have said in a public lecture that women were innately less suited to high positions in science. My own view is that the field of sex differences in mind needs to proceed in a fashion that is sensitive to this history of conflict by cautiously looking at the evidence and being careful not to overstate what can be concluded. What we will see in this lecture is that whilst males as a group show a stronger interest in systems, which is at the heart of science, when it comes to individuals, there are great female scientists and there are great male scientists. Once again, the evidence says nothing about individuals. As we will see, the data actually require us to look at each individual on their own merits, as individuals may or may not be typical for their sex.

Systemizing and empathizing

'Empathizing' is the drive to identify another person's emotions and thoughts and to respond to these with an appropriate emotion. Empathizing allows you to *predict* a person's behaviour and to care about how others feel. In this essay, I review the evidence that, in general, females spontaneously empathize to a greater degree than do males. 'Systemizing' is the drive to analyze the variables in a system in order to derive the underlying rules that govern its behaviour. Systemizing also refers to the drive to construct systems. Systemizing allows one to *predict* the behaviour of a system and to control it. I review the evidence that, on average, males spontaneously systemize to a greater degree than do females.[1]

---

[1] S. Baron-Cohen, S. Wheelwright, J. Lawson, *et al.*, 'The exact mind: empathising and systemising in autism spectrum conditions', in: U. Goswami (ed.), *Handbook of Cognitive Development*. Oxford: Blackwell, 2002.

Table 2.1: *Main types of analyzable systems*

- **Technical** systems (e.g., a computer, a musical instrument, a hammer)
- **Natural** systems (e.g., a tide, a weather front, a plant)
- **Abstract** systems (e.g., mathematics, a computer program, syntax)
- **Social** systems (e.g., a political election, a legal system, a business)
- **Organizable** systems (e.g., a taxonomy, a collection, a library)
- **Motoric** systems (e.g., a sports technique, a performance, a musical technique)

Empathizing is close enough to the standard English definition to need little introduction, and I will come back to it shortly. But systemizing is a new concept and needs a little more definition. By a 'system' I mean something that takes inputs and delivers outputs. To systemize, one uses 'if-then' (correlation) rules. The brain zooms in on a detail or parameter of the system and observes how this varies. That is, it treats a feature of a particular object or event as a variable. Alternately, a person actively, or systematically, manipulates a given variable. One notes the effect(s) of operating on one single input in terms of its effects elsewhere in the system (the output). The key data structure used in systemizing is [input-operation-output]. If I do x, a changes to b. If z occurs, p changes to q. Systemizing therefore requires an exact eye for detail.

There are at least six kinds of systems that the human brain can analyze or construct, as shown in Table 2.1. Systemizing is an inductive process. One watches what happens each time, gathering data about an event from repeated sampling, often quantifying differences in some variables within the event and observing their correlation with variation in outcome. After confirming a reliable pattern of association – that is, generating predictable results – one forms a rule about how a particular aspect of the system works. When an exception occurs, the rule is refined or revised. Otherwise, the rule is retained. Systemizing works for phenomena that are ultimately lawful, finite and deterministic. The explanation is exact, and its truth-value is testable. ('The light went on because the switch was in the down position.') Systemizing is of almost no use for predicting moment-to-moment changes in a person's behaviour. To predict human behaviour, empathizing is required. Systemizing and empathizing are wholly different kinds of processes.

Empathizing involves the attribution of mental states to others and involves an appropriate affective response to the other's affective state. It includes not only what is sometimes called 'theory of mind', or mentalizing,[2] but also encompasses the common English words 'empathy' and 'sympathy'. Although systemizing and empathizing are in one way similar because they are processes that allow us to make sense of events and make reliable predictions, they are in another way almost the opposite of each other. Empathizing involves an imaginative leap in the dark in the absence of complete data. ('Maybe she didn't phone me because she was feeling hurt by my comment.') The causal explanation is at best a 'maybe', and its truth may never be provable. Systemizing is our most powerful way of understanding and predicting the law-governed inanimate universe. Empathizing is our most powerful way to understand and predict the social world. Ultimately, empathizing and systemizing depend on separate, independent regions in the human brain.

The main brain types

In this essay I will argue that systemizing and empathizing are two key dimensions that define the male and female brain. We all have both systemizing and empathizing skills. One can envisage five broad types of brain, as Table 2.2 shows. This essay concerns itself primarily with those on the extreme male brain end of the spectrum. Individuals who have this psychological profile may be talented systemizers, but they are often, at the same time, 'mind-blind'.[3] The evidence reviewed here suggests that not all men have the male brain and not all women have the female brain. Expressed differently, some women have the male brain, and some men have the female brain. My central claim here is only that *more* males than females have a brain of type S, and *more* females than males have a brain of type E. I will review the evidence

[2] J. Morton, A. Leslie and U. Frith, 'The cognitive basis of a biological disorder: autism', *New Scientist* **14**, 1995, 434–8.
[3] S. Baron-Cohen, *Mindblindness: An Essay on Autism and Theory of Mind.* Boston, MA: MIT Press, 1995.

Table 2.2: *The main brain types*

Profile	Shorthand Equation	Type of Brain
Individuals in whom empathizing is more developed than systemizing.	E>S	'female' (or Type E)
Individuals in whom systemizing is more developed than empathizing.	S>E	'male' (or Type S)
Individuals in whom systemizing and empathizing are both equally developed.	S=E	'balanced' (or Type B)
Individuals in whom systemizing is hyperdeveloped while empathizing is hypodeveloped (the autistic end of the spectrum). They may be talented systemizers, but at the same time, they may be 'mind blind'.	S>>E	extreme male brain
Individuals who have hyperdeveloped empathizing skills, while their systemizing is hypodeveloped – they may be 'system blind'.	E>>S	extreme female brain (postulated)

supporting these profiles. In the final section of this essay, I will high-light the role of culture and biology in these sex differences.

The female brain: empathizing

What is the evidence for female superiority in empathizing? In the studies summarized here, sex differences of a small but statistically significant magnitude have been found.

- *Sharing and turn taking.* On average, girls show more concern for fairness, while boys share less. In one study, boys showed fifty times greater competition, as compared to girls, while girls showed twenty times greater turn taking, as compared to boys.[4]

---

[4] W. R. Charlesworth and C. Dzur, 'Gender comparisons of preschoolers' behavior and resource utilization in group problem solving', *Child Development* **58**, 1987, 191–200.

- *Rough and tumble play or 'rough housing'* (wrestling, mock fighting, etc.). Boys show more of this than do girls. Although such activity is often playful, it can hurt or be intrusive. Lower empathizing levels are necessary to engage in rough and tumble play.[5]
- *Responding empathically to the distress of other people.* Girls from the age of one year show greater concern for others through sad looks, sympathetic vocalizations and comforting as compared to boys. Also, more women than men report frequently sharing the emotional distress of their friends. Women also show more comforting, even to strangers, than men do.[6]
- *Using a 'theory of mind'.* As early as three years of age, little girls are ahead of boys in their ability to infer what people might be thinking or intending.[7]
- *Sensitivity to facial expressions.* Women are better at decoding nonverbal communication, picking up subtle nuances from tone of voice or facial expression, or judging a person's character.[8]
- *Empathy.* Women score higher than men on questionnaires designed to measure empathic response.[9]
- *Values in relationships.* More women than men value the development of altruistic, reciprocal relationships, which by definition require empathizing. In contrast, more men value power, politics and competition.[10] Girls are more likely to endorse cooperative items on a questionnaire and to rate the establishment of intimacy as more important than the establishment of dominance. In contrast, boys are more likely than girls to endorse competitive items and to rate social status as more important than intimacy.[11]
- *Disorders of empathy.* Disorders such as psychopathic personality disorder or conduct disorder are far more common among males.[12]

---

[5] E. E. Maccoby, *The Two Sexes: Growing Up Apart, Coming Together.* Cambridge, MA: Belknap Press/Harvard University Press, 1998.

[6] M. L. Hoffman, 'Sex differences in empathy and related behaviors', *Psychological Bulletin* **84**, 1977, 712–22.

[7] F. G. Happe, 'The role of age and verbal ability in the theory of mind task performance of subjects with autism', *Child Development* **66**, 1995, 843–55.

[8] J. A. Hall, 'Gender effects in decoding nonverbal cues', *Psychological Bulletin* **85**, 1978, 845–58.

[9] M. H. Davis, *Empathy: A Social Psychological Approach.* Ed. J. Harvey, Brown & Benchmark Social Psychology Series. Boulder, CO: Westview Press, 1994.

[10] A. Ahlgren and D. W. Johnson, 'Sex differences in cooperative and competitive attitudes from the 2nd through the 12th grades', *Developmental Psychology* **15**, 1979, 45–9.

[11] G. P. Knight and C.-C. Chao, 'Gender differences in the cooperative, competitive, and individualistic social values of children', *Motivation and Emotion* **13**, 1989, 125–41.

[12] K. A. Dodge. 'Social cognition and children's aggressive behavior', *Child Development* **51**, 1980, 162–70; R. J. Blair, 'A cognitive developmental approach to morality: investigating the psychopath', *Cognition* **57**, 1995, 1–29.

- *Aggression*. Even in normal quantities, this can only occur with reduced empathizing. Here again, there is a clear sex difference. Males tend to show far more 'direct' aggression (pushing, hitting, punching, etc.), while females tend to show more 'indirect' (relational, covert) aggression (gossip, exclusion, cutting remarks, etc.). Direct aggression may require an even lower level of empathy than indirect aggression. Indirect aggression needs better mind-reading skills than does direct aggression because its impact is strategic.[13]
- *Murder.* This is the ultimate example of a lack of empathy. Daly and Wilson analyzed homicide records dating back over 700 years from a range of different societies. They found that 'male-on-male' homicide was thirty to forty times more frequent than 'female-on-female' homicide.[14]
- *Establishing a 'dominance hierarchy'*. Males are quicker to establish such hierarchies. This in part reflects their lower empathizing skills because often a hierarchy is established by one person pushing others around to become the leader.[15]
- *Language style*. Girls' speech is more cooperative, reciprocal and collaborative. In concrete terms, this is also reflected in girls being able to continue a conversational exchange with a partner for a longer period. When girls disagree, they are more likely to express their different opinion sensitively, in the form of a question rather than an assertion. Boys' talk is more 'single-voiced discourse'; that is, the speaker presents only his own perspective. The female speech style is more 'double-voiced discourse'; girls spend more time negotiating with their partner, trying to take the other person's wishes into account.[16]
- *Talk about emotions*. Women's conversations involve much more talk about feelings, while men's conversations tend to be more object- or activity-focused.[17]
- *Parenting style*. Fathers are less likely than mothers to hold their infants in a face-to-face position. Mothers are more likely to follow through the child's choice of topic in play, while fathers are more likely to impose their own topic. Also, mothers fine-tune their speech more often to match their children's understanding.[18]

[13] N. R. Crick and J. K. Grotpeter, 'Relational aggression, gender, and social-psychological adjustment', *Child Development* **66**, 1995, 710–22.

[14] M. Daly and M. Wilson, *Homicide.* New York: Aldine de Gruyter, 1988.

[15] F. F. Strayer, 'Child ethology and the study of preschool social relations', in: H. C. Foot, A. J. Chapman and J. R. Smith (eds.), *Friendship and Social Relations in Children.* Chichester: John Wiley & Sons, 1980.

[16] P. M. Smith, *Language, the Sexes, and Society.* Oxford: Blackwell, 1985.

[17] D. Tannen, *You Just Don't Understand: Women and Men in Conversation.* New York: William Morrow & Company, 1990.

[18] T. G. Power, 'Mother- and father-infant play: a developmental analysis', *Child Development* **56**, 1985, 1514–24.

- *Face preference and eye contact.* From birth, females look longer at faces, particularly at people's eyes, whereas males are more likely to look at inanimate objects.[19]

Females have also been shown to have better language ability than males. It seems likely that good empathizing would promote language development[20] and vice versa, so these factors may not be independent.

The male brain: systemizing

The relevant domains to explore for evidence of systemizing include any fields that are in principle rule-governed. Thus, chess and football are good examples of systems, but faces and conversations are not. As noted previously, systemizing involves monitoring three elements: input, operation and output. The operation is what was done or what happened to the input in order to produce the output. What is the evidence for a stronger drive to systemize in males?

- *Toy preferences.* Boys are more interested than girls in toy vehicles, weapons, building blocks and mechanical toys, all of which are open to being 'systemized'.[21]
- *Adult occupational choices.* Some occupations are almost entirely male. These include metal-working, weapon-making, musical instrument-making, and the construction industries, such as boat-building. The focus of these occupations is on creating systems.[22]
- *Maths, physics and engineering.* These disciplines all require high systemizing and are largely male-dominated. The Scholastic Aptitude Math Test (SAT-M) is the mathematics part of the test administered nationally to college applicants in the United States. Males on average score fifty points higher than females on this test.[23] Considering only

[19] J. Connellan, S. Baron-Cohen, S. Wheelwright, *et al.*, 'Sex differences in human neonatal social perception', *Infant Behavior and Development* 23, 2000, 113–18.
[20] S. Baron-Cohen, D. A. Baldwin and M. Crowson, 'Do children with autism use the speaker's direction of gaze strategy to crack the code of language?' *Child Development* 68, 1997, 48–57.
[21] K. D. Jennings, 'People versus object orientation in preschool children: do sex differences really occur?' *Journal of Genetic Psychology* 131, 1977, 65–73.
[22] D. C. Geary, *Male, Female: The Evolution of Human Sex Differences.* Washington, DC: American Psychological Association, 1998.
[23] C. P. Benbow, 'Sex differences in mathematical reasoning ability in intellectually talented preadolescents: their nature, effects, and possible causes', *Behavioral and Brain Sciences* 11, 1988, 169–232.

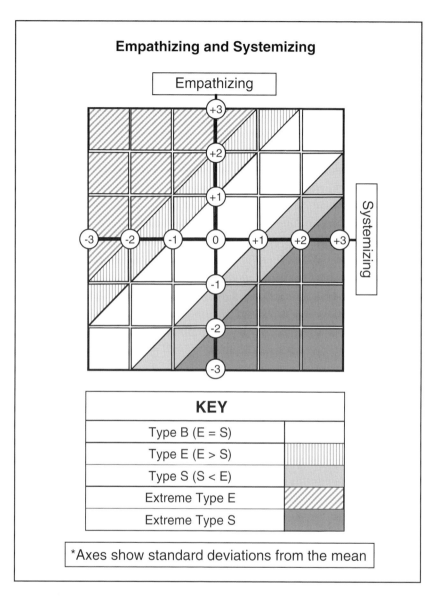

FIGURE 2.1: Schematic diagram of different types of human brain in relation to their empathizing and systemizing abilities. See Table 2.2 for further elaboration of these types.

individuals who score above 700, the sex ratio is 13:1 (men to women).[24]

- *Constructional abilities*. On average men score higher than women in an assembly task in which people are asked to put together a three-dimensional (3–D) mechanical apparatus. Boys are also better at constructing block buildings from two-dimensional blueprints. Lego bricks can be combined and recombined into an infinite number of systems. Boys show more interest than girls in playing with Lego. Boys as young as three years of age are also faster at copying 3–D models of outsized Lego pieces. Older boys, from the age of nine years, are better than girls at imagining what a 3–D object will look like if it is laid out flat. Boys are also better at constructing a 3–D structure from just an aerial and frontal view in a picture.[25]

- *The Water Level Task*. Originally devised by the Swiss child psychologist Jean Piaget, the Water Level Task involves a bottle that is tipped at an angle. Individuals are asked to predict the water level. Women more often draw the water level aligned with the tilt of the bottle and not horizontal, as is correct.[26]

- *The Rod and Frame Test*. If a person's judgement of vertical is influenced by the tilt of the frame, he or she is said to be 'field dependent'; that is, their judgement is easily swayed by extraneous input in the surrounding context. If they are not influenced by the tilt of the frame, they are said to be 'field independent'. Most studies indicate that females are more field dependent; i.e., women are relatively more distracted by contextual cues, and they tend not to consider each variable within a system separately. They are more likely than men to state erroneously that a rod is upright if it is aligned with its frame.[27]

- *Good attention to relevant detail*. This is a general feature of systemizing and is clearly a necessary part of it. Attention to relevant detail is superior in males. One measure of this is the Embedded Figures Test. On average, males are quicker and more accurate in locating a target object from a larger, complex pattern.[28] Males, on average, are also better at detecting a particular feature (static or moving) than are women.[29]

[24] D. C. Geary, 'Sexual selection and sex differences in mathematical abilities', *Behavioral and Brain Sciences* **19**, 1996, 229–84.

[25] D. Kimura, *Sex and Cognition*. Cambridge, MA: MIT Press, 1999.

[26] M. A. Wittig and M. J. Allen, 'Measurement of adult performance on Piaget's water horizontality task', *Intelligence* **8**, 1984, 305–13.

[27] H. A. Witkin, H. B. Lewis, M. Hertzman, *et al.*, *Personality Through Perception*. New York: Harper & Brothers, 1954.

[28] R. Elliot, 'Interrelationship among measures of field dependence, ability, and personality traits', *Journal of Abnormal and Social Psychology* **63**, 1961, 27–36.

[29] D. Voyer, S. Voyer and M. Bryden, 'Magnitude of sex differences in spatial abilities: a meta-analysis and consideration of critical variables', *Psychological Bulletin* **117**, 1995, 250–70.

- *The Mental Rotation Test.* This test provides another example in which males are quicker and more accurate. This test involves systemizing because it is necessary to treat each feature in a display as a variable that can be transformed (e.g., rotated) and then predict the output, or how it will appear after transformation.[30]
- *Reading maps.* This is another everyday test of systemizing, because features from 3–D input must be transformed to a two-dimensional representation. In general, boys perform at a higher level than girls in map reading. Men can also learn a route by looking at a map in fewer trials than women, and they are more successful at correctly recalling greater detail about direction and distance. This observation suggests that men treat features in the map as variables that can be transformed into three dimensions. When children are asked to make a map of an area that they have only visited once, boys' maps have a more accurate layout of the features in the environment. More of the girls' maps make serious errors in the location of important landmarks. Boys tend to emphasize routes or roads, whereas girls tend to emphasize specific landmarks (the corner shop, the park, etc.). These strategies of using directional cues versus using landmark cues have been widely studied. The directional strategy represents an approach to understanding space as a geometric system. Similarly, the focus on roads or routes is an example of considering space in terms of another system, in this case a transportation system.[31]
- *Motoric systems.* When people are asked to throw or catch moving objects (target- directed tasks), such as playing darts or intercepting balls flung from a launcher, males tend to perform better than females. In addition, on average men are more accurate than women in their ability to judge which of two moving objects is travelling faster.[32]
- *Organizable systems.* People in the Aguaruna tribe of northern Peru were asked to classify a hundred or more examples of local specimens into related species. Men's classification systems included more sub-categories (i.e., they introduced greater differentiation) and were more consistent among individuals. Interestingly, the criteria that the Aguaruna men used to decide which animals belonged together more closely resembled the taxonomic criteria used by Western (mostly

---

[30] D. W. Collins and D. Kimura, 'A large sex difference on a two-dimensional mental rotation task', *Behavioral Neuroscience* **111**, 1997, 845–9.

[31] L. A. M. Galea and D. Kimura, 'Sex differences in route-learning', *Personality and Individual Differences* **14**, 1993, 53–65.

[32] W. Schiff and R. Oldak, 'Accuracy of judging time to arrival: effects of modality, trajectory, and gender', *Journal of Experimental Psychology: Human Perception and Performance* **16**, 1990, 303–16.

male) biologists.[33] Classification and organization involves systemizing because categories are predictive. With more fine-grained categories, a system will provide more accurate predictions.

- *The Systemizing Quotient.* This is a questionnaire that has been tested among adults in the general population. It includes forty items that ask about a subject's level of interest in a range of different systems that exist in the environment, including technical, abstract and natural systems. Males score higher than females on this measure.[34]
- *Mechanics.* The Physical Prediction Questionnaire (PPQ) is based on an established method for selecting applicants to study engineering. The task involves predicting which direction levers will move when an internal mechanism of cog wheels and pulleys is engaged. Men score significantly higher on this test, compared with women.[35]

Culture and biology

At age one year, boys strongly prefer to watch a video of cars going past, an example of predictable mechanical systems, than to watch a film showing a human face. Little girls show the opposite preference. Young girls also demonstrate more eye contact than do boys at age one year.[36] Some investigators argue that, even by this age, socialization may have caused these sex differences. Although evidence exists for differential socialization contributing to sex differences, this is unlikely to be a sufficient explanation. Connellan and colleagues showed that among one-day-old babies, boys look longer at a mechanical mobile, which is a system with predictable laws of motion, than at a person's face, an object that is next to impossible to systemize. One-day-old girls show the opposite profile.[37] These sex differences are therefore present very early in life.

[33] S. Atran, 'Core domains versus scientific theories: evidence from systematics and Itzaj-Maya folkbiology', in: L. A. Hirschfeld and S. A. Gelman (eds.), *Mapping the Mind: Domain Specificity in Cognition and Culture.* Cambridge: Cambridge University Press, 1994.

[34] S. Baron-Cohen, J. Richler, D. Bisarya, *et al.*, 'The Systemising Quotient (SQ): An investigation of adults with Asperger Syndrome or high functioning autism and normal sex differences', *Philosophical Transactions of the Royal Society, Series B.* **358**, 2003, 361–74.

[35] J. Lawson, S. Baron-Cohen and S. Wheelwright, 'Empathizing and systematizing in adults with and without Asperger Syndrome', *Journal of Autism and Developmental Disorders* **34**, 2004, 301–10.

[36] S. Lutchmaya and S. Baron-Cohen, 'Human sex differences in social and nonsocial looking preferences at 12 months of age', *Infant Behavior and Development.* In press.

[37] See n. 19.

This raises the possibility that, while culture and socialization may partly determine the development of a male brain with a stronger interest in systems or a female brain with a stronger interest in empathy, biology may also partly determine this. There is ample evidence to support both cultural determinism and biological determinism.[38] For example, the amount of time a one-year-old child maintains eye contact is inversely related to the prenatal level of testosterone.[39] The evidence for the biological basis of sex differences in the mind is reviewed elsewhere.[40]

Autism: an extreme form of the male brain

Autism is diagnosed when a person shows abnormalities in social development and communication and displays unusually strong obsessional interests from an early age.[41] Asperger Syndrome (AS) has been proposed as a variant of autism. It is seen in children who have normal or high IQ scores and who develop speech at the normal developmental age. Today, approximately one in 200 children have one of the 'autistic spectrum conditions', which include AS.[42] Autism spectrum conditions are far more common in males than females. Among individuals with high-functioning autism (HFA) or AS, at least ten males are affected for every one female. These conditions are also strongly heritable[43] and neurodevelopmental in origin. Considerable evidence supports structural and functional differences in certain regions of the brain. For example, the amygdala is abnormal in size in many individuals with autistic spectrum disorders, and it may not respond in the usual fashion to cues of emotional expression.[44]

[38] A. H. Eagly, *Sex Differences in Social Behavior: A Social-Role Interpretation*. Hillsdale, NJ: Lawrence Erlbaum Associates, 1987; C. Gouchie and D. Kimura, 'The relationship between testosterone levels and cognitive ability patterns', *Psychoneuroendocrinology* **16**, 1991, 323–34.

[39] S. Lutchmaya, S. Baron-Cohen and P. Raggatt, 'Foetal testosterone and eye contact in 12 month-old infants', *Infant Behavior and Development* **25**, 2002, 327–35.

[40] S. Baron-Cohen, *The Essential Difference: Men, Women, and the Extreme Male Brain*. New York: Basic Books, 2003.

[41] Task Force on DSM-IV. *Diagnostic and Statistical Manual of Mental Disorders*. 4th edn. Washington, DC: American Psychiatric Association, 1994.

[42] U. Frith (ed.), *Autism and Asperger Syndrome*. Cambridge: Cambridge University Press, 1991.

[43] A. Bailey, P. Bolton and M. Rutter, 'A full genome screen for autism with evidence for linkage to a region on chromosome 7q', *Human Molecular Genetics* **7**, 1998, 571–8.

[44] S. Baron-Cohen, H. A. Ring, E. T. Bullmore, *et al.*, 'The amygdala theory of autism', *Neuroscience and Biobehavioral Reviews* **24**, 2000, 355–64.

The extreme male brain (EMB) theory of autism was first informally suggested by Hans Asperger in 1944. According to the 1991 translation by Uta Frith, he wrote, 'The autistic personality is an extreme variant of male intelligence. Even within the normal variation, we find typical sex differences in intelligence. . . . In the autistic individual, the male pattern is exaggerated to the extreme.'[45] In 1997 this controversial hypothesis was re-examined.[46] We can now test the EMB theory empirically, as the female brain (E>S), the male brain (S>E), and the balanced brain (E=S) have been defined. According to the EMB theory, people with autism or AS should always fall in the dark-grey zone as illustrated in Figure 2.1.

Evidence for the extreme male brain theory

Initial tests are beginning to provide positive proof of this theory.[47] A number of studies utilizing different approaches and standard instruments indicate that people with autism show markedly impaired empathizing. Some of the convergent lines of evidence are summarized here.

- *Mind reading.* Girls score better than boys on standard 'theory of mind' tests, and children with autism or AS tend to perform even worse than do normal boys.[48] Children with autism have specific delays and difficulties in the development of 'mind reading', and they are unable to make sense of or predict another's feelings, thoughts and behaviour. Autism has been referred to as a condition of 'mindblindness'.[49]
- *The Empathy Quotient.* On this questionnaire, females score higher than males, and people with AS or HFA score even lower than males.[50]

[45] See n. 42.
[46] S. Baron-Cohen and J. Hammer, 'Is autism an extreme form of the "male brain"?' in: C. Rovee-Collier and L. P. Lipsitt (eds.), *Advances in Infancy Research.* Vol. 11. Greenwich, Conn: Ablex Publishing, 1997.
[47] S. Baron-Cohen, S. Wheelwright, V. Stone, *et al.*, 'A mathematician, a physicist, and a computer scientist with Asperger syndrome: performance on folk psychology and folk physics test', *Neurocase* 5, 1999, 475–83; S. Baron-Cohen, 'The cognitive neuroscience of autism: evolutionary approaches', in: M. S. Gazzaniga (ed.-in-chief), L. Cosmides and J. Tooby (eds.), *The New Cognitive Neurosciences.* 2nd edn. Cambridge, MA: MIT Press; 2000.
[48] See n. 7.
[49] See n. 3.
[50] See n. 34.

sarcastic   **stern**

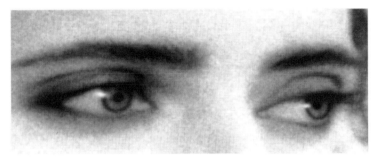

suspicious   **dispirited**

preoccupied   **grateful**

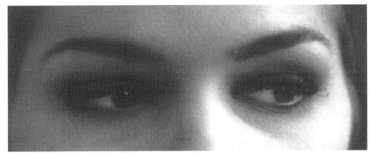

insisting   **imploring**

FIGURE 2.2: The 'Reading the Mind in the Eyes' Test. Subjects are asked to select the most fitting of the four words to match the expression in the eyes.

- *The 'Reading the Mind in the Eyes' Test* (Figure 2.2). Females score higher on this test than males, but people with AS do not even score as well as males.[51]
- *The Complex Facial Expressions Test*. Similar to the other tests mentioned, females score higher than males, and people with AS score even lower than do males.[52]

[51] S. Baron-Cohen, T. Jolliffe, C. Mortimore, *et al.*, 'Another advanced test of theory of mind: evidence from very high functioning adults with autism or Asperger syndrome', *Journal of Child Psychology and Psychiatry and Allied Disciplines* **38**, 1997, 813–22.

[52] S. Baron-Cohen, S. Wheelwright and T. Jolliffe, 'Is there a "language of the eyes"? Evidence from normal adults and adults with autism or Asperger syndrome', *Visual Cognition* **4**, 1997, 311–31.

- *Eye contact.* Females make eye contact more often and maintain it for longer periods of time than do males. People with autism or AS make less eye contact than males.[53]
- *Language development.* Girls develop vocabulary faster than boys, and children with autism are even slower than males to develop vocabulary.[54]
- *Pragmatics.* Females tend to be superior to males at chatting with others and at understanding the pragmatics of conversation. It is precisely this aspect of language which people with AS find most difficult.[55]
- *The Faux-Pas Test.* Females are better than males at judging what would be socially insensitive or potentially hurtful or offensive. People with autism or AS have even lower scores on tests of this than do males.[56]
- *The Friendship Questionnaire (FQ).* This instrument assesses empathic styles of relationships. Women score higher than males on this questionnaire, and adults with AS score even lower than do normal males on the FQ.[57]

There also exists a growing body of evidence that supports the superior systemizing abilities of individuals with autism spectrum disorders.

- *Islets of ability.* Some people with autism spectrum disorders have 'islets of ability' in mathematical calculation, calendrical calculation, syntax acquisition, music, or precise memory for railway timetable information.[58] For high-functioning individuals, this can lead to considerable achievement in mathematics, chess, mechanical knowledge, and other factual, scientific, technical or rule-based subjects. All of these areas are highly systemizable domains, and most are also domains that are more interesting to males than to females in the general population.

[53] See n. 39; J. Swettenham, S. Baron-Cohen, T. Charman, *et al.*, 'The frequency and distribution of spontaneous attention shifts between social and nonsocial stimuli in autistic, typically developing, and nonautistic developmentally delayed infants', *Journal of Child Psychology and Psychiatry and Allied Disciplines* **39**, 1998, 747–53.

[54] S. Lutchmaya, S. Baron-Cohen and P. Raggatt, 'Foetal testosterone and vocabulary size in 18- and 24-month-old infants', *Infant Behavior and Development* **24**, 2002, 418–24.

[55] S. Baron-Cohen, 'Social and pragmatic deficits in autism: cognitive or affective?' *Journal of Autism and Developmental Disorders* **18**, 1988, 379–402.

[56] S. Baron-Cohen, M. O'Riordan, V. Stone, *et al.*, 'Recognition of faux pas by normally developing children and children with Asperger syndrome or high-functioning autism', *Journal of Autism and Developmental Disorders* **29**, 1999, 407–18.

[57] S. Baron-Cohen and S. Wheelwright, 'The friendship and relationship questionnaire', *Journal of Autism and Developmental Disorders* **33**, 2003, 509–17.

[58] S. Baron-Cohen and P. Bolton, *Autism: The Facts.* Oxford: Oxford University Press, 1993.

- *Attention to detail.* People with autism also tend to pay extra-fine attention to detail. For example, on the Embedded Figures Test (EFT) males score higher than females, and people with AS or HFA score even higher than males. This is not a systemizing test *per se*, but it is a measure of detailed local perception, which is a prerequisite for successful systemizing.[59] On visual search tasks, males demonstrate better attention to detail than do females, and people with autism or AS have even faster, more accurate visual search skills.[60]
- *Preference for rule-based, structured, factual information.* People with autism are strongly drawn to structured, factual and rule-based information. A male bias for this kind of information is also found in the general population.
- *Intuitive physics.* Males score higher than females on tests of intuitive physics. People with AS tend to score higher than males on such tests.[61]
- *Toy preference.* In general, boys prefer construction-type and vehicle toys more than girls do. Clinical reports suggest that children with autism or AS demonstrate a very strong preference towards these types of toys as well.[62]
- *Collecting.* Boys engage in more collecting or organizing of items than girls, and people with autism show this characteristic to an even greater extent.[63]
- *Obsessions with closed systems.* Most individuals with autism are naturally drawn to predictable things, such as computers. Unlike people, computers follow strict laws. Computers are closed systems; that is, all the variables are well defined within the system, and they are knowable, predictable and, in principle, controllable. Other individuals with autism may not make computers their target of understanding but may latch on to a different, equally closed system, such as bird migration or trainspotting.[64]

---

[59] T. Jolliffe and S. Baron-Cohen, 'Are people with autism or Asperger syndrome faster than normal on the Embedded Figures Test?' *Journal of Child Psychology and Psychiatry and Allied Disciplines* **38**, 1997, 527–34.

[60] M. O'Riordan, K. C. Plaisted, J. Driver, *et al.*, 'Superior visual search in autism', *Journal of Experimental Psychology: Human Perception and Performance* **27**, 2001, 719–30.

[61] S. Baron-Cohen, S. Wheelwright, V. Scahill, *et al.*, 'Are intuitive physics and intuitive psychology independent? A test with children with Asperger syndrome', *Journal of Developmental and Learning Disorders* **5**, 2001, 47–78.

[62] See n. 35.

[63] See n. 40.

[64] S. Baron-Cohen and S Wheelwright, '"Obsessions" in children with autism or Asperger syndrome: a content analysis in terms of core domains of cognition', *British Journal of Psychiatry* **175**, 1999, 484–90.

- *The Systemizing Quotient.* As noted previously in this essay, males score higher on this test, and people with autism and AS score even higher than normal males on this instrument.[65]

Finally, some evidence rooted in biology and genetics supports the EMB theory of autism.

- *The Autism Spectrum Quotient (the AQ).* Males in the general population score higher than females, and people with AS or HFA score highest of all on this instrument.[66]
- *Sexually dimorphic somatic markers.* Finger-length ratio is a sexually dimorphic somatic marker. In general, males tend to have a longer ring finger compared to their second finger, which is different than the ratio in females. People with autism or AS show an even greater difference in the ratio of ring-finger to second- finger length.[67]
- *Puberty.* Males with autism are reported to show precocious puberty, which correlates with increased levels of circulating testosterone.[68]
- *Familiality of talent.* Males are over-represented in occupations such as engineering, which require good systemizing but where a mild impairment in empathizing is not necessarily an impediment to success.[69] There is a higher rate of autism in the families of those talented in fields such as mathematics, physics, and engineering, as compared to those who are most talented in the humanities.[70] These findings suggest that the extreme male cognitive style is, in part, inherited.

Conclusions and future research

The above evidence suggests that the male brain is characterized by type S (where S>E), the female brain by type E (where E>S), and the autistic brain is an extreme of the male brain (S>>E). Referring to

---

[65] See n. 34.
[66] S. Baron-Cohen, S. Wheelwright, R. Skinner, *et al.*, 'The autism-spectrum quotient (AQ): evidence from Asperger syndrome/high-functioning autism, males and females, scientists and mathematicians', *Journal of Autism and Developmental Disorders* **31**, 2001, 5–17 [published correction appears in *Journal of Autism and Developmental Disorders* **31**, 2001, 603].
[67] J. T. Manning, S. Baron-Cohen, S. Wheelwright, *et al.*, 'The 2nd to 4th digit ratio and autism', *Developmental Medicine and Child Neurology* **43**, 2001, 160–4.
[68] S. Tordjman, P. Ferrari, V. Sulmont, *et al.*, 'Androgenic activity in autism', *American Journal of Psychiatry* **154**, 1997, 1626–7.
[69] S. Baron-Cohen, S. Wheelwright, C. Stott, *et al.*, 'Is there a link between engineering and autism?' *Autism* **1**, 1997, 101–8.
[70] S. Baron-Cohen, P. Bolton, S. Wheelwright, *et al.*, 'Does autism occur more often in families of physicists, engineers, and mathematicians?' *Autism* **2**, 1998, 296–301.

Figure 2.1, development of an autism spectrum condition indicates that an individual's brain type is shifted towards the lower-right quadrant. For males, it is a small degree of shift, from type S to extreme type S. For females, the movement is greater, from type E to extreme type S. The causes of this shift remain unclear, but candidate factors include both genetic differences and prenatal testosterone levels.[71]

The model in Figure 2.1 predicts that the extreme female brain (EFB) exists. How would such individuals behave? By definition, their brain type is in the upper-left quadrant of Figure 2.1. Their ability to empathize is significantly better than other people in the general population, but their systemizing abilities are impaired. This category would include people who have difficulty understanding mathematics, physics, mechanical objects, chemistry and the like as systems[72] but who are extremely accurate at tuning in to others' feelings and thoughts. Would such a profile carry with it any disability? A person with EFB would be 'system-blind'. In our society, there remains considerable tolerance for such individuals. It is hoped that people who are 'mind-blind' will also enjoy the same tolerance by society.

We know something about the neural circuitry of empathizing,[73] but at present we know very little about the neural circuitry of systemizing. Research will hopefully begin to reveal the key brain regions involved in systems processing.

Acknowledgements

The Medical Research Council (UK) supported my work during the writing of this essay: I am grateful to Sally Wheelwright and Johnny Lawson for the development of Figure 2.1. Parts of this essay are based on work published elsewhere (S. Baron-Cohen, 'The extreme male brain theory of autism', *Trends in Cognitive Science* **6**, 2002, 248–54).

---

[71] See n. 43.
[72] See n. 1.
[73] S. Baron-Cohen, H. A. Ring, S. Wheelwright, *et al.*, 'Social intelligence in the normal and autistic brain: an fMRI study', *European Journal of Neuroscience* **11**, 1999, 1891–8.

FURTHER READING

S. Baron-Cohen, *The Essential Difference: Men, Women and the Extreme Male Brain*. Harmondsworth/London/New York: Penguin/Allen Lane/ Basic Books, 2003.

D. Geary, *Male, Female: The Evolution of Human Sex Differences*. Washington, DC, and London: American Psychological Association, 1998.

D. Kimura, *Sex and Cognition*. Cambridge, MA: MIT Press, 1997.

3 Why apes and humans kill

RICHARD WRANGHAM

Department of Anthropology, Harvard University

Most animals rarely kill their own. Among non-human primates, for example, many species live in social groups and fight with their neighbours. Battles involve loud calls, displays, chases and occasional grapples – just enough for the larger or stronger group to convince its opponent to retreat. Yet although inter-group contests may be intense and frequent, and although they may determine access to critical resources, they hardly ever lead to death. For most primates the goal is to win or to retreat without being damaged.

But humans often aim to kill. So why is our species different? Why do we regularly and deliberately kill our own kind?

At the simplest level there are two kinds of answer: killing is either unnatural or natural. The contrasting positions are encapsulated by the ideas of Jean-Jacques Rousseau and Thomas Hobbes, who have become icons for opposing views of human nature. For Rousseau, we are an essentially peaceful species. For Hobbes, war is our natural state. As we will see, both positions can be criticized.

Rousseau's thinking was dominated by the concept of the 'noble savage'. He was so impressed by reports of people's spontaneous tendency to care for each other that he claimed conflict was absent in a state of nature. 'With passions so tame', he wrote, 'and so salutary a curb, men, rather wild than wicked, and more attentive to guard against mischief than to do any to other animals, were not exposed to any dangerous dissensions'.[1]

---

[1] J.-J. Rousseau, *A discourse upon the origin and the foundation of the inequality among mankind* (1762), in: C. W. Eliot (ed.), *French and English Philosophers: Descartes, Rousseau, Voltaire, Hobbes*, New York: P. F. Collier, 1938, 165–228 (here 190).

To followers of Rousseau, only unnatural features can explain killing by humans. Candidate traits include our invention of weapons, which make killing so easy; our development of brutal ideologies; the intrusion of evil into a good world; or perhaps the acquisition of wealth, harems or farmland, rewards so desirable that they justify intensely risky contests. The essential notion is that killing is foreign to our nature. Our biology does not promote our violence.

Hobbes, by contrast, thought human nature responsible for violence. He found ready evidence in the English Civil War. When a man travels, he said: 'he arms himself and seeks to go well accompanied; when going to sleep, he locks his doors; when even in his house, he locks his chests; and this when he knows there be laws and public officers armed to revenge all injuries shall be done him. . . . Does he not there as much accuse mankind by his actions as I do by my words?' Thinking that his own experience of homeland insecurity captured the essence of human nature, he characterized society as a 'war of all against all'.[2] Followers of Hobbes vary in which aspects of human nature they think most responsible for violence. For Immanuel Kant it was patriotism: 'War . . . seems to be ingrained in human nature, and even to be regarded as something noble to which man is inspired by his love of honour, without selfish motives'. For one of Fyodor Dostoyevsky's characters, it was lust: 'It's just their defencelessness that tempts the tormentor, just the angelic confidence of the child who has no refuge and no appeal, that sets his vile blood on fire. In every man, of course, a beast lies hidden – the beast of rage, the beast of lustful heat at the screams of the tortured victim, the beast of lawlessness let off the chain. . .'. Either way, killing is natural. So there is no need to focus on weapons as precipitating factors, when merely by using coalitions to beat each other up we can kill. Ideology becomes less important than our species' psychology as a cause of violence. Anyone can kill, not just psychopaths. We do so not because we're evil, but because opportunity leads us to take advantage of our killing power.[3]

[2] T. Hobbes, *Of Man, being the first part of Leviathan*, in: Eliot (ed.), *French and English Philosophers*, 311–417 (here 389).

[3] I. Kant, *Perpetual Peace: A Philosophical Essay* (1795), New York: Garland, 1972; F. Dostoyevsky, *The Brothers Karamazov*, New York: Knopf, 1992, ii, V, 4.

The debate between followers of Rousseau and Hobbes has continued for some 300 years. At first glance, the fact that animals rarely kill each other might seem to support Rousseau's position. Indeed, it has often been claimed that animals never kill each other at all, a notion that would strongly support Rousseau. Even as recently as the 1960s, the great ethologist Konrad Lorenz claimed that animals are inhibited from killing each other. Lorenz was an acute observer and there was little wrong in his reports that dominant wolves of the same pack were inhibited from killing subordinates (as long as they made the appropriate signals of appeasement). But Lorenz over-generalized. Although wild wolves are inhibited from attacking members of their own groups, they often kill members of neighbouring groups. Contrary to what we might hope and what once seemed to be the case, killing is not restricted to humans. It can happen naturally.

For too long, claims about animals have been based on imagination more than fact. It is only in the last two or three decades that behavioural biology has been rooted in sound empirical data from the wild. We now know that across species, neither the Rousseauian nor the Hobbesian extreme is right. Animals are neither universal pacifists nor generally violent. Instead, the killing tendency varies widely across species.

Perhaps the most dramatic primate example of a non-violent species is the muriqui or woolly spider monkey of South America. Karen Strier has observed groups of muriqui for more than two decades, and has long wanted to explore the role of dominance and aggression among males. But she has found it impossible. After years of trying, she has been unable to detect dominance relationships among males. The problem is that her subjects are so insistently uncompetitive. Even when a female is sexually interested and several males are with her, the males are models of courtesy. They can line up on a branch waiting to take their turns to mate, without even a hint of impatience. They hardly ever fight.

Chimpanzees were once thought to have an almost equally pacific social life, but that period of supposed innocence lasted only to the time when researchers began following well-habituated chimpanzees throughout their community ranges. Observers then saw violence so extreme that it led to deaths. A variety of other species are now known to be capable of the same level of violence.

The new understanding is grisly. But it is helpful because it presents the opportunity to assess why, in a few selected species, natural selection favours killing. If there are evolutionary rules, we can deduce them; and we can then consider what they mean for humans. Does our species follow the same principles of evolutionary biology that generates killing in chimpanzees or other animals? Or does human killing have its own independent logic?

Is chimpanzee killing natural or unnatural?

Modern study of behaviour among wild chimpanzees began in the 1960s in Gombe National Park, Tanzania, when Jane Goodall persuaded a community to tolerate her presence at close quarters. She quickly learned the basics of their lives. Every individual belongs to a social community, but travels alone or in small parties according to choice. She named her main study group the Kasekela community. It had some fifty members who roamed over four main valleys in their daily search for fruits. She was unaware, at first, that the edges of the community range were defended.

But by the early 1970s researchers observing individual Kasekela chimpanzees all day had accumulated regular sightings of tense encounters with strangers at the edge of the community range. Mostly the interactions were mere display. Parties with several males typically met in stand-off. They retreated from each other after they had displayed with loud calls and occasional chases. These group battles were harmless.

But occasionally inter-community interactions had serious consequences. First an infant was killed. Then, in 1974, observers watched in horror as a party of Kasekela males hunted down and caught an adult male from a neighbouring group, the Kahama community. They beat this male so badly that after a few days he died.

During the next four years similar attacks were repeated. By 1977, the Kahama community was gone. All its males had disappeared, as many as seven killed by the unprovoked hunts of their Kasekela neighbours.

This startling series of events was followed by evidence of similar violence in a second population of chimpanzees. Toshisada Nishida was also working in western Tanzania, and like Jane Goodall had habituated two wild chimpanzee communities. He worked in the Mahale Mountains, where a small community (K-group) lived next to a large community (M-group). The two communities battled every year for access to an area of K-group's range that produced a short-lived crop of excellent fruits, and M-group always won. Most battles were non-violent, but Nishida and his colleagues also found that K-group males were disappearing under suspicious circumstances – such as after M-group males had rushed off and called as if they were fighting, in an area where a dead K-group male was later found. Over a period of ten years, five males vanished in their prime. K-group followed the fate of the Kahama community. Their community was eliminated. Nishida concluded that several, and maybe all, of the K-group males had been killed by the M-group males.

Killing of adult strangers is now known from two other communities, both in Uganda's Kibale National Park. It has also been recorded when captive chimpanzees were introduced to a wild area containing chimpanzees. A recent report from the Congo concluded that without emergency veterinary intervention, almost half of the males released into the natural forest would have been killed by wild chimpanzees. Even in captivity, males have combined to attack and kill a rival.

Set against this litany of carnage are two long-term chimpanzee study sites where no inter-community killings have been recorded. But even those are not sites of peace. At one of them (Taï, in the Ivory Coast), the chimpanzees attempt to isolate victims and attack them in the classic manner, though they have not yet been seen to succeed. At another (Budongo), where the borders are still poorly known, males have killed both infants from neighbouring communities and an adult male in their own community.

From the five long-term sites where chimpanzees live with neighbours we now have more than 180 years of data on eleven communities. The five sites yield direct evidence of kills of neighbours in thirty-three cases, and a further sixteen suspected victims. They also show repeated interest in killing, evidenced from patrolling of borders to attempts to isolate, immobilize and intensely batter or cut a victim.

Two conclusions emerge. First, rates of brutality vary between sites and over time. The next generation of study will identify factors responsible for this variation. They seem likely to include factors affecting the stability and size of parties, such as the intensity of competition or the degree of clumping in the food supply. Conceivably there are also cultural differences among chimpanzee populations in their propensity for such aggression.

But second, chimpanzee violence is a regular phenomenon that occurs in many situations. It's in the nature of chimpanzees to kill.

Why do chimpanzees kill?

Given that wild chimpanzees sometimes deliberately kill each other they offer an excellent opportunity to understand how natural selection can favour the killing tendency. The fact that chimpanzees are one of human's two closest relatives makes this a particularly interesting question, since it raises the possibility that the same explanation applies to ourselves.

We could approach the question 'Why do chimpanzees kill each other?' in several different ways. We might ask about how the behaviour develops during an individual's life, or what precipitating factors set off the psychological or physiological chain that ends in a vicious beating, or whether we can trace the behaviour to parallel patterns in species that are close evolutionary relatives. But here I focus at the adaptive level partly because we know most about it, and partly because it is a useful starting-point for understanding the other approaches. My aim is to assess the costs and benefits that push chimpanzees towards lethal violence.

There are several possibilities. In theory the aggression might be carried out by a few sociopathic apes. Or all chimpanzees could have an innate drive to kill strangers. Alternatively, if all chimpanzees are potential killers they might be driven to it by shortage of food, or by the lure of reward in the form of new females, or by the prospect of access to especially valuable food areas.

In fact, however, none of these ideas fits the current evidence, as I describe below. But there's another hypothesis that fits the data well.

FIGURE 3.1: Victim of inter-group conflict: Rasputin, a chimpanzee
slaughtered in the wild in Tanzania (© Martin N. Muller)

This is the idea that killing pays because it reduces the number of
rivals in the rival community.

One of the clues that points us to this answer comes from the sensitivity of male chimpanzees to the number of their immediate compan-
ions. In Kibale we find that the more males that a party includes, the
more likely they are to interact with their neighbours. There are two
obvious reasons.

The first is relevant to the kinds of uninhibited punishment that
cause death. Those intense beatings happen when a group of at least
three individuals, and usually more, are able to isolate a rival on his
own. On those occasions the victim is sometimes surrounded and
repeatedly hit with blows stronger than any human can make.
Alternatively the victim may be immobilized by being held down by his
wrists and ankles. In that position he is helpless. Martin Muller's
photograph of a Northern community male that was trapped in this
way by eight Kanyawara males in September 1998 shows the problem
(Figure 3.1). The Northern male had cuts and bruises on much of his

ventral surface. His thorax was torn across, and both testicles had been removed. He died on the spot, though the next day his killers returned to drag his body about. Even so, there was hardly a scratch on his back surface. The only exception was telling testimony to the cruelty of mobbing chimpanzees. On the victim's left elbow his skin had been torn from the front and ripped off at the back as well. It's a striking example of just how deliberately chimpanzees damage their helpless rivals. Just as has been seen in Gombe, an attacker had apparently clamped his front teeth onto the victim's skin, then reared his head back to tear the skin away.

The effect of a massive imbalance of power is clear. When a sufficiently large party of males surrounds a lone rival, it is easy for them to inflict lethal damage. Yet the attackers are hardly ever hurt. The wounding is quick and safe. This suggests that chimpanzees' decisions about when to approach strangers would be influenced partly by whether many males are travelling together.

Experiments conducted by Michael Wilson richly confirm that idea, When males hear a tape-recording that purports to be a lone stranger's call, their willingness to call depends on the number of males in the party that hears the call. A group of females, or a lone male, is predictably silent. Their response rate rises sharply with the number of males, so that when four or more males are together, they call more than 90 per cent of the time. An equivalent effect applies to their decision to travel towards the call, or for the speed at which they do so, or for the distance that they advance. More males together give confidence to each.

Such finely tuned assessment serves the aggressors well in decisions about attacking loners. It also helps them succeed in battles, when opposing parties each contain several males. Unlike the brutal beatings of surprised loners, battles involve little physical contact. Instead, chimpanzees on either side display at each other with bluff charges and screams. Encounters between well-balanced sides can continue for thirty minutes or more at a time. Battles sometimes end in both parties retiring in opposite directions to feed, but more usually the side with fewer males ends up retreating.

A lost battle can have important consequences. While the losers live to fight another day, they may nevertheless avoid the contested area for weeks. As a result they feed less well than they otherwise would have done. In Kibale one of the study community's most productive fruit-groves lies in the south of its range, and in recent years the community has left the area immediately after its annual defeats in battle by a large neighbouring community. Likewise, every year in Mahale, the large M-group community would supplant the smaller K-group community from the valuable fruit-crop ripening in the shared area of their range. In Gombe the long-term consequences of such competition have been revealed by an analysis of eighteen years of data during which the Kasekela community's territory oscillated from small to large. When the community was squeezed into an area as small as five square kilometres, feeding competition was relatively intense: travelling parties were small, chimpanzees had low body weight, the interval between births averaged six years, and infant survival was low. The good times occurred when the community was strong enough to defend a territory almost three times the size. Mothers then increased their output of babies to one every four years, and all other measures of their condition also improved.

Such evidence shows that a big territory is valuable, but it still doesn't explain why chimpanzees kill to defend it. Of the many hypotheses that have been suggested, current data now suggest that most can be eliminated. For example in theory, chimpanzees might have a generalized tendency to kill: but in fact they select their victims and their times to attack with precision. In theory, some males might be sociopathic killers: but all males participate roughly equally. In theory, killing might be a response to competition: but there's no indication that it happens more when resources are in short supply – more likely, it happens when food is abundant. In theory, killing might have been directed to obtaining females, or might have been a result of a shortage of mating opportunities. But the killers don't get immediate matings or even (normally) any immigrating females as a result. Nor is there any short-term benefit in the form of access to contested food supplies.

In contrast to those ideas, there is one explanation that fits the data well. This is the hypothesis that chimpanzees kill rivals whenever they can do so safely, because killing raises the likelihood of winning future battles. Certainly chimpanzees kill specifically when they have a low risk of being hurt themselves. Furthermore they direct their attacks almost entirely to males, the sex that alone defends the territory. Thus while female strangers are often found alone and may be chased and beaten, more than 90 per cent of adult victims of lethal inter-community aggression are male. Even infants killed by adults are mostly male. Those who die, therefore, are mainly the present and future defenders of the territory.

By killing rival males when it is cheap to do so, chimpanzees shift the balance of power between their own community and the neighbours. Since males are the defenders of the territory and are recruited only from within, deaths of males in the neighbouring community have a long-term effect in weakening its power to win battles. They therefore increase the chance that the killers will later win access to larger areas; and larger areas mean more food, and hence more babies. This means that an important benefit of killing is its likely impact on reproduction in the ensuing years. Following the harsh logic of natural selection, deliberate killing of males looks beneficial because it undermines the rivals' defensive capacity, and increases the relative power of the killers.[4]

Predatory aggression as a strategy

A striking feature of this system is that it doesn't fit the traditional view of natural violence as resulting from escalated contest. That conventional concept sees fatal aggression as resulting from a loser's failure to respond to a high likelihood of losing. It implies that the victim of violence is the weakling that over-estimates his chances. The corpse is of the fighter that stayed too long in the battle.

But the pattern of violence among chimpanzees allows losers no choice about how long to fight. Rather than making poor decisions in the fight, chimpanzee victims are taken by surprise and attacked until they are fatally wounded. They are victims of circumstance rather than

---

[4] J. M. Williams, G. Oehert and A. E. Pusey, 'Why do male chimpanzees defend a group range', *Animal Behaviour* **68**, 2004, 523–32.

of failed assessment, and the violence they experience is not so much contest as predation.

The predatory nature of these attacks is clear from the fact that they are often not provoked by any immediate contact with the neighbouring community. They can begin when a large party of well-fed chimpanzees walk to their territorial edge to conduct a border patrol, during which they might spend an hour or more looking for neighbours to stalk, hunt and attack. In the event of a successful attack there is no immediate pay-off other than the satisfaction the aggressors experience from the act itself. The implication is that natural selection has favoured in chimpanzees a tendency to relish the prospect and performance of such brutality.

While this is an unpleasant concept it is not restricted to chimpanzees. This kind of violence is a particular example of offensive aggression, provoked merely by the opportunity to kill. Among social carnivores, such as lions, killing can occur much as in chimpanzees, with several individuals surrounding an isolated rival so as to prevent its escape, and then carefully wounding it so badly that it dies. A different form of predatory killing occurs in species such as the hanuman langur monkey or savanna baboon. There, adult males take opportunities to kill infants that are not related to them. In certain species of fig-wasp, the first activity of newly emerged males is to search for males about to hatch, and kill them. Rates of killing can be high. Among wolves, 50 per cent or more of adult deaths have been attributed to kills by other wolves. In some monkeys, similar numbers of infants die at the hands and teeth of adult males. In each of these types of killing, there is a different kind of advantage to the killer. But there is a common feature nevertheless: the killing is easy.

The distribution of these various kinds of offensive aggression, or predatory killing, is not predictable from evolutionary relationship. For example while chimpanzees are killers, they have a sister species, the bonobo, that is equally closely related to humans as chimpanzees yet shows no signs of predatory killing. Bonobos have not been recorded killing members of their own species, and although they relish meat, they are even unknown to kill monkeys (a universal habit of chimpanzees). Nor do they show any evidence of attempts to search for isolated rivals, commit violent infanticide, or carry out severe beatings. Why bonobos should be so different from chimpanzees is a fascinating question. The best guess points to their evolutionary ecology. Bonobos nowadays travel in parties that are more stable than those of chimpanzees, a luxury allowed them by the distribution of their unique foods: unlike chimpanzees, their habitats appear to include more widely distributed 'meadows' of food on the forest floor, a food supply that allows bonobos to forage in stable parties. Possibly, therefore, a difference in food supply underlies a difference in the stability of parties, which in the case of bonobos means that individuals are hardly ever forced to travel alone. Among chimpanzees, by contrast, when food is scarce, parties are forced to break up, leaving lone individuals vulnerable to attack.

Rather than evolutionary relationship, the common thread running through the species that exhibit predatory aggression is the same as explains the timing of chimpanzee violence: at the time of the attack, the killers are predictably safe. By virtue of being in larger parties, or bigger, or more mobile than their victims, aggressors can kill while experiencing a very low risk of being hurt themselves. If these are circumstances that applied throughout long stretches of evolutionary history, natural selection appears to favour the psychology that underlies predatory killing – whether in chimpanzees, lions, hanuman langurs or fig-wasps.

As with the evolution of competitive relationships more generally, there is an irony about these systems of fatal aggression. Even though the tendency to kill is beneficial for the killers it is damaging for the species. For example, if there were no killing, the species as a whole would experience a lower death rate. But natural selection is blind to such effects. Evolution has no mechanism for getting rid of behavioural propensities merely because they have undesirable effects at the level of the species. It takes a change in evolutionary ecology, or else a conscious mind, to achieve that goal.

War in hunter-gatherers

If chimpanzee violence is a system for using a temporary advantage to achieve a long-term power gain, the obvious hypothesis is that human inter-group killing follows the same principle. I now consider this possibility.

Ultimately the question is whether chimpanzees and humans share overlapping psychological propensities for behaviours that lead to success at killing enemies. Since this kind of problem concerns our evolutionary traits it depends on our behaviour when we lived as hunter-gatherers – the predominant subsistence style of human evolution. Hunting-and-gathering (or nomadic foraging) probably originated in a simple form about 2 million years ago. It flourished in diverse ways throughout the human world, and started giving way to agriculture only within the last 15,000 years.

Essentially no archaeological evidence exists concerning hunter-gatherer collective violence of the type practiced by chimpanzees, in which individuals are picked off one at a time. But some hunter-gatherers still lived as 'hunters in a world of hunters' even when first described by outsiders, and these people offer a glimpse of traditional practices of war. Scattered around the globe from the Arctic to Tasmania and Tierra del Fuego, such peoples typically belonged to tribes of a few hundred or thousand people, living in small residential camps averaging less than fifty people. Tribes had territories that they defended against neighbours, although some peoples such as desert Australians allowed regular trespass. Relations between tribes, and styles of war, varied between regions sufficiently to force us to be cautious about overall patterns. But some reasonably confident generalizations have nevertheless emerged.

First, it is important to recognize that there are large differences between chimpanzee collective violence and hunter-gatherer war. Chimpanzee violence is very simple. Among other things it lacks weapons, complex intention, political language, symbolic behaviour and ideology. It also involves clashes between single communities, rather than among several groups facing each other in alliance as sometimes occurred among foragers.

Despite such differences, however, some important commonalities are found between our two species. To begin with, in contrast to a widespread perception that hunter-gatherers tended to live in peace, there was typically a state of hostility between neighbouring tribes. As a result, war deaths were frequent. In a recent review, for example, anthropologist Azar Gat found that 'all the available evidence indicates very high killing rates among all known simple hunter-gatherer societies'. Not every anthropologist reaches such an unambiguous conclusion, but all agree that war was widespread and regularly dangerous.[5]

Killers were almost invariably men but the circumstances of killing otherwise varied widely. At one extreme were opportunistic massacres of vulnerable strangers found by surprise; at another, there were deliberate raids in search of victims in far-away camps. But whatever the style, the great majority of deaths occurred when the killers had an overwhelming advantage, whether of surprise, power or technology. As Radcliffe-Brown described of people in the Andaman Islands, the point was that lethal aggression depended on feeling safe: 'The whole art of fighting was to come upon your enemies by surprise, kill one or two of them and then retreat. . . . They would not venture to attack the enemy's camp unless they were certain of taking it by surprise. . . . If they met with any serious resistance or lost one of their own number, they would immediately retire . . . the aim of the attacking party was to kill the men. . . .'.[6]

Similar principles appear to have been at work among slash-and-burn farmers living in politically independent communities, such as the Yanamamö of South America or the Dugum Dani of highland New Guinea. Raiding often exacted a high death toll. By contrast serious battles were rare. Battles were largely ceremonial, responsible for few deaths. The big danger incurred by New Guinea battles appears to have been that occasionally, one of the opponents would discover that its fighting force was alarmingly small compared to that of a rival alliance. When that happened, a ritual display could turn into a massacre.

These trends indicate that to a large extent, hunter-gatherer or slash-and-burn warfare included the same essential features that make violence among chimpanzees both unusual and explicable. In these societies participation in warfare was voluntary, but since attack-

---

[5] A. Gat, 'The human motivational complex: evolutionary theory and the causes of hunter-gatherer fighting, part II: Proximate, subordinate, and derivative causes', *Anthropological Quarterly* **73**, 2000, 74–88.

[6] A. R. Radcliffe-Brown, *The Andaman Islanders: A Study in Social Anthropology*, Cambridge: Cambridge University Press, 1922.

ers were rarely killed it was not considered especially dangerous. By taking advantage of temporary imbalances of power, men safely killed rivals from neighbouring communities. It seems clear that an accumulation of deaths could have a lasting impact on power relations between neighbours.

As mentioned above, there are clearly important differences between chimpanzee and human violence. For example, power-imbalances among people do not depend merely on a large group surprising a lone individual. They can also be influenced by weaponry, tactics and deceit: guests invited to a feast might be slaughtered, for instance. Consider also the way in which the two species judge 'safe killing'. There is no evidence that chimpanzees do more than assess the immediate chance of success. Among hunter-gatherers, by contrast, warriors appear to assess the safety of an attack not merely by whether they can kill their victims, but also by how effectively the victims' allies may be able to strike back. Such complexities doubtless alter the frequency and pattern of violence.

Despite such important differences, the similarities in inter-group killing between the two species are striking because they contain the essential elements needed to explain how selection favours killing in chimpanzees. In both cases, the killing is overwhelmingly predatory, meaning that it is safe for the killers because the victims are attacked by surprise when they are especially vulnerable. Attacks are directed particularly towards males. And in both cases, group defence of territory appears important for achieving reproductive success. The similarities suggest the hypothesis that the psychology of adult males is adapted in both cases to taking advantage of helpless rivals by killing them, because of the long-term advantages of being militarily superior to the neighbours. The specific brain adaptations are a matter for speculation, but an obvious possibility is that predatory killing activates a neural reward system. Certainly there is no indication that in either species, adult males suffer psychologically from killing in these circumstances. Adult males show considerable enthusiasm at opportunities to attack, and in small-scale human societies they can receive explicit cultural or material rewards for their success.

Wars in state society

The hypothesis that humans and chimpanzees have important similarities in their tendencies for predatory violence (and its functional effects) does not imply that inter-state war should follow the same principles. In state societies warriors are part of a military hierarchy, and therefore have little freedom of action. When sufficiently dominated by their leaders, they can be forced into being violent in ways that threaten their own survival. In particular, they may be forced into participating in a lethal battle, i.e., an escalated contest between coalitions in which opponents risk being killed while attempting to kill. Lethal battles among humans are an evolutionary novelty among mammals. They appear to have arisen with state-level societies. A strict hierarchy appears to be an important pre-condition for their occurrence. Lethal battles present a challenging problem for evolutionary analysis.

However, even in state-level societies much inter-group violence happens in well-organized predatory attacks. The art of guerrillas, for example, is essentially to surprise the enemy rather than be forced into battle. Such tactics are prominent in wars of decolonialism or national resistance.

Even within lethal battles predatory killing has a role. Sometimes, such as in the Battle of Agincourt (1415), small groups of soldiers reportedly succeed by finding isolated members of the enemy and using their temporary advantage to dispatch the enemy individually. Again, often the major killing in battles happens not when opposing sides face each other, but after one army loses its nerve, turns and runs: the victors can then kill easily and safely. The same is true of many massacres, when one side holds its enemies captive. As in chimpanzees, these are cases where the killers are rarely hurt. They therefore make obvious sense.

But in addition to predatory killing, lethal battles are also widespread in state warfare. In contrast to predatory killing in simpler warfare, much evidence suggests that lethal battles do not excite the warriors. Soldiers ordered to charge an opposing army are typically reluctant to do so, and most soldiers are reported to avoid killing when possible. Those who kill often find the experience traumatic. Soldiers generally fight not from excitement but in a state of fear – fearful of letting down their close colleagues, of being punished by their officers, or of being killed before they destroy the enemy's power. Far from being spontaneously confident and enthusiastic, soldiers going into battle routinely gain false courage by being given alcohol.

Even the behaviour of military leaders is quite unlike that of predatory killers. Unlike chimpanzees or warriors in small-scale societies, leaders consistently suffer from positive illusions: they ordinarily deceive themselves into exaggerating their probabilities of victory. A classic example is John F. Kennedy's failure in the Bay of Pigs fiasco, when the USA supported a force that in retrospect was patently inadequate and destined to be defeated. 'How could I have been so stupid as to let them proceed?' asked Kennedy afterwards. Many a losing general has asked the same question. A result of this predictable intelligence failure is that battles happen more often, and more intensely, than they would under perfect knowledge. Why positive illusions occur predictably among military leaders is a fascinating question, but whatever the reason, their occurrence stresses that these interactions are markedly different from lethal violence among chimpanzees and hunter-gatherers. We might sometimes wish that our generals were as effective at assessing the odds as chimpanzees are. That way, they could avoid many unnecessary battles.

In sum, lethal battles among humans are a cultural novelty not found in other mammals. The unique forms of conflict that they include must be explained by reference to specifically human features, such as the ability of leaders to order men into battle and the assessment failures of leaders in escalated conflicts. The biological propensities resulting from a putative evolutionary history of collective violence therefore have little relevance to explaining the killing behaviour of soldiers in confronting armies.

Resolution

If lethal battles are not explicable by reference to chimpanzee violence, the question that remains is whether the tendency for safe killing among humans, in its various guises from hunter-gatherer warfare to guerrilla attacks or even deliberate massacres, is importantly

influenced in our own species by an evolutionary history of chimpan-
zee-like predatory violence. There are two obvious kinds of answer:
one cautious, and the other, in my opinion, more likely to be helpful.

The cautious or weak inference is that predatory violence by humans
has its own unique neural basis which has no significant relationship
to the psychology of chimpanzees. Many people who favour this conclusion would see human violence as entirely culturally influenced, and
would therefore describe the parallels to chimpanzee violence as
curious but meaningless coincidences. Accordingly, the fact that
human predatory violence arises easily in rather consistent forms
resembling those of chimpanzees, from street gangs and small-scale
societies to civil wars and genocides, would remain an unsolved evolutionary and cultural puzzle. This inference would leave uncertain the
extent to which predatory killing among humans can be predicted. For
instance, it would explain the difference in killing rates between men
and women in terms of elusively defined cultural influences.

Unfortunately, some people may be tempted to favour this conclusion for reasons largely unrelated to the evidence. For example, they
might equate a biological explanation with a determinist philosophy.
Because they dislike the implication that a biological explanation con-
demns us to violence, therefore, they might arbitrarily reject the notion
that biological predispositions could influence human violence. This
line of thinking is sometimes found among social scientists who are
rightly scornful of simplistic genetic or eugenic accounts of behaviour.

But there are at least two reasons why such a rejection would be
unfortunate. First, it misreads the philosophical implications of behavioural biology. Contrary to the implication of the phrase 'biological
determinism', biological analysis does not predict behaviour on the
basis of individual differences in genetic make-up. For example, as this
essay has stressed, the conditions in which chimpanzees attack rivals
are determined by assessment rather than by propensities for violence.
Whether a particular attack occurs, therefore, depends not on the
genetic constitution of the participants but on circumstances affecting
the power imbalances, such as the patterns of food availability and the
size, composition and social dynamics of neighbouring communities.
In short, although a biological analysis explains the background to vio-

lence, we still need a sociological, ecological and cultural account of local conditions to explain specific patterns. In this sense, biology is not determinist in the way that it is sometimes accused of being. Fear of biological determinism would therefore be a wrong reason for rejecting the possibility that violence among humans and chimpanzees has been subject to selection for similar neural mechanisms.

Second, rejection of a biological analysis would close the door on potentially fruitful hypotheses, such as emerge from considering the alternative, or 'strong', inference. This is that similar biological propensities underlie predatory violence in both chimpanzees and humans. According to this idea, the tendency to kill safely has been favoured by natural selection and likely involves parallel neural reward systems. It would mean that humans, especially males, are expected to take advantage of perceived power imbalances to kill their rivals. It therefore draws attention to the mechanisms of assessment, and how people judge safety, as being important influences on the control of violent behaviour.

Data are needed to settle the relative worth of the weak and strong inferences. Ideally we would be informed by species comparisons of genetic, neural and psychological mechanisms. Meanwhile, based on the growing evidence of similarities between chimpanzee and human biology, the ease and consistency with which humans evince predatory killing, and the evidence that human evolutionary ecology offered the appropriate selective context, my intuition favours the strong inference. Furthermore, from the perspective of looking for a more peaceful world, although it might be less cautious academically it has the advantage of being the more cautious in practice. It suggests that Lord Acton was biologically accurate when he wrote that 'Power tends to corrupt; and absolute power corrupts absolutely'.[7] If we take humans to be fundamentally similar to chimpanzees in their attitude to violence, we should be highly alert to the dangers inherent in major acquisitions of power. The biological intuition warns us that men will continue to look for opportunities to massacre their rivals, and that we

---

[7] Lord Acton, *Historical Essays and Studies*, ed. John Figgis and Reginald V. Lawrence, London: Macmillan, 1907, 504.

should never drop our guard. The bad news is that we have to work to stop men from uniting to kill their rivals. The good news is that if we maintain balances of power, we can expect peace.

FURTHER READING

L. H. Keeley, *War Before Civilization*. New York: Oxford University Press, 1996.

R. C. Kelly, *Warless Societies and the Origins of War*. Ann Arbor, MI: University of Michigan Press, 2000.

R. L. Kelly, *The Foraging Spectrum: Diversity in Hunter-gatherer Lifeways*. Washington, DC: Smithsonian Institution, 1995.

S. A. LeBlanc, *Constant Battles*. New York: St. Martin's Press, 2003.

H. H. Turney-High, *Primitive War: Its Practice and Concepts*. Columbia, SC: University of South Carolina Press, 1949.

R. W. Wrangham and D. Peterson, *Demonic Males: Apes and the Origins of Human Violence*. Boston: Houghton Mifflin, 1996.

4 The roots of warfare

BARRY CUNLIFFE

Institute of Archaeology, University of Oxford

In the opening sequence of his 1968 film *2001 – a space odyssey*, Stanley Kubrick offers a stunning image of the beginning of warfare when one of his ape men picks up a long bone from a decaying animal carcass and uses it to bludgeon an opponent to death. In Kubrick's brilliant orchestration of this moment there is a palpable sense of awe and wonder among the creatures as their minds come to terms with the magnitude of the discovery. To Kubrick, it is a defining moment – the beginning of man's progress to civilization. This is a stark, uncomfortable, message which we need to examine in some detail.

Enter the philosophers

The nature of human aggression has long fascinated philosophers. One of the first to confront the problem head on was the British philosopher Thomas Hobbes (1588–1679). Oxford-trained, he lived through, and survived, the Civil War, an experience which cannot have failed to have had an impact on his thinking. Politically he walked a difficult tight-rope. He was a firm believer in strong government, and therefore supported the King against Parliament, but he was not prepared to accept the divine right of kings. His basic philosophy was that of a materialist who believed that rational explanations could be found for all human behaviour.

In his famous work *Leviathan*, published in English in 1651, he addresses the question of 'the state of nature'. All men desire what is good for themselves and everyone aspires to peace so long as it is in the interests of the individual. But if danger is perceived then it is reasonable to adopt any means to secure self-preservation – this represents the 'rights of nature'. It follows that if the 'rights of nature' are

63

enforced the 'state of nature' will result in which the life of man will inevitably be, in that famous and much-quoted phrase, 'solitary, poor, nasty, brutish and short'. If the aim of humanity is lasting preservation ('the goal of reason'), it is necessary to create a stable society which requires the individual to give up the rights of nature and to invest decision-making in a sovereign individual or group able to act for the long-term good. Only by individuals agreeing 'covenants' with that central authority can a state of perpetual war be averted.

Thus, what Hobbes is arguing is not that man is naturally violent or biologically conditioned to dominate, but that the condition of war is a social condition which can only be controlled or averted through 'covenants'. Individually humans desire 'good' but have the propensity to achieve it through violence, unless constrained by agreed conventions.

A rather different view was taken by the French philosopher Jean-Jacques Rousseau (1712–78). Unlike Hobbes, he believed that 'nothing could be more gentle' than man in his natural state. This 'noble savage' lived in a golden age that 'mankind was formed ever to remain in'. What went wrong was that humans organized themselves in societies with unnatural laws, such as safeguarding private property and upholding monogamy, which inevitably led to conflict. In *Social Contract*, published in 1762, he writes of the sovereignty of the entire citizen body over itself creating a general will which would naturally promote liberty and equality and both would arise from, and further encourage, fraternity. His views were influential among the leaders of the French Revolution but it is a mercy for him that he did not live long enough to see the results of their political application.

At the time that Rousseau was writing, travellers were bringing back stories of their encounters with primitive peoples in various parts of the world. Some of their observations were decidedly disconcerting for him and he showed a distinct reluctance to incorporate the newly available empirical evidence into his philosophy. When told of an encounter between previously uncontacted aboriginal Tasmanians and a group of unarmed French emissaries, who had removed all clothing for the occasion – an encounter which resulted in a violent attack on the Frenchmen – he responded, 'Is it possible that good Children of

Nature can really be so wicked?' (thus demonstrating a typical academic's ploy – when in difficulty, organize a discussion group).

The oppositions between the Hobbes and Rousseau views can be summarized in this way: for Hobbes, man has the natural capacity for violence but is constrained by social convention and strong leadership; for Rousseau, man is by nature non-violent but is led to violence by unnatural social constraints.

Throughout the nineteenth century, a century of imperialism and colonization, the views of Hobbes were widely favoured not least since they provided a comforting rationale to underpin the west European desire to extend the comforts of Christianity and civilization to those perceived to be in need of them. Moreover, events in Europe and encounters with native peoples abroad served to strengthen the view that not only was Rousseau's 'social contract' a seriously flawed concept but his belief in the 'noble savage' was dangerously naïve.

The contribution of the sciences

While philosophers could debate the nature of man and ethnographers could introduce a flood of observational detail, it was not until 1798, with the publication of Malthus' *An Essay on the Principle of Population*, that a scientific theory capable of providing an explanatory model was introduced. Thomas Malthus (1766–1834) published his work, known as the *First Essay*, when he was a Fellow of Jesus College, Cambridge, essentially to debunk what he considered to be the wildly Utopian visions of the French Revolution. It was an exuberant polemic which he reworked and greatly improved, publishing a new text – the *Second Essay* – in 1803.

Malthus' thesis was elegantly simple and rests on two basic principles. The first was the observation that unchecked population growth always exceeds the growth of the means of subsistence. All animals have the capacity to increase their population in geometrical progression and therefore, if unchecked, that population will double every twenty-five years. This greatly outstrips means of subsistence which can only increase arithmetically. His second observation is that actual population growth is kept in balance with available food supply by two kinds of checks: *positive checks* such as warfare, disease, starvation and infanticide, all of which increase death rate; and *preventative checks* such as marriage laws and sexual prohibitions, which lower birth rate.

If we translate this into simple modern demography, we can say that the natural tendency for population to increase exponentially is always constrained by the holding capacity of the environment. When a population reaches or exceeds the holding capacity, the society goes into a state of stress. This can be relieved in a variety of ways, by increasing the holding capacity, for example by the introduction of new crops or technologies, or by moving a sector of the population into a new ecological niche (i.e., by outward colonization). But while these controls are brought into use, society experiences a heightened level of aggression which, almost inevitably, involves warfare.

To Thomas Malthus, then, warfare among human groups was the unavoidable result of the natural laws of population growth. This view underlies the 'ecological approach' to the study of culture which was developed in the late twentieth century by scholars like Julian Steward, who refers to it as 'cultural ecology', or Marvin Harris, who writes of 'cultural materialism'. To both, the ecological approach is a form of functionalism. Thus, for them, the function of warfare is to act as a mechanism which adjusts population size and in doing so rectifies ecological imbalances. We will return to these issues again below.

Malthus' *Essay* was to have a profound effect on another of the key figures in the debate, Charles Darwin (1809–82). Darwin writes in his autobiography how, as early as 1838,

> I happened to read for amusement Malthus on *Population*, and
> being well prepared to appreciate the struggle for existence which
> everywhere goes on, from long-continued observation of the habits
> of animals and plants, it at once struck me that under these
> circumstances favourable variations would tend to be preserved and
> unfavourable ones to be destroyed. The result would be a new species.
> Here, then, I had at last got a theory by which to work.
>
> C. Darwin, *Autobiography*, ed. N. Barlow, London: Collins, 1958, 120

This theory of natural selection was to form the core of his *Origin of Species*, published in 1859, and the *Descent of Man*, which appeared in 1871. The Malthus/Darwin view, then, accepted that not only was aggression and warfare 'natural', but that they were positive forces for change since it was the 'fittest', or the most able, who would survive. To put it another way, the process of natural selection favoured the characteristic of aggression.

Anthropologists and stateless societies

The second half of the twentieth century has seen an upsurge of interest in the study of warfare largely as the result of a number of detailed observational studies carried out in various parts of the world among contemporary stateless societies. We have already referred to the 'ecological approach' of scholars like Julian Steward and Marvin Harris, who saw the function of warfare as a mechanism to control population size through increasing the death rate and to force periodic readjustments in the allocation of ecological resources. This view is closely similar to the Malthusian approach. Other scholars, like Konrad Lorenz, took a different view, arguing that the aggressive drive is innate and spontaneous and not the outcome of environmental stimulation. An individual male of any given species will readily display his aggression, though he is constrained through social forces to direct it outside the group thus helping to define the integrity of his own community in relation to others. Robert Ardrey was sympathetic to this analysis, arguing that by nature man was a territorial animal programmed to threaten, and, if need be, to fight in defence of his territory. For him, too, aggression is a basic drive. The Lorenz/Ardrey view – that aggression is innate and beneficial, and by implication is selected for – is not far removed from the Darwinian approach. It need hardly be said here that many anthropologists disagreed with all this and a lively debate ensued.

The generalizations of anthropological theorists are, of course, based on a wealth of observational detail, much of it coming from studies of stateless societies living in non-developed parts of the world. This rich and varied database has been trawled by a number of scholars intent on producing cross-cultural studies of primitive warfare. Two of the most recent, Lawrence Keeley's *War before Civilization* (1996) and Steven LeBlanc's *Constant Battles* (2003), both stress the extent and intensity of warfare throughout the world. Both writers are adamant in their belief that warfare is endemic and the notion of the peaceful savage is a myth.

Two classic studies serve to illustrate the point – the Yanomamö of the rainforests of Venezuela and Brazil and the Maring of the highlands of New Guinea.

The Yanomamö are known largely through the fieldwork of Napoleon Chagnon in the 1960s. They live in small groups, each of about a hundred, in fortified villages surrounded by their fields. They farm plantains, bananas and manioc, and hunt to supplement the diet with animal protein. Population densities are low, so there is no obvious ecological pressure, and indeed the villages and fields are usually surrounded by large buffer zones up to thirty miles wide and yet warfare is endemic. Raids and ambushes are the norm while the decimation of entire village communities, involving the massacre of the men and children and the capture of the women, is not infrequent. One vivid account by a twelve-year-old captured by a raiding party in the 1930s tells how, after their menfolk had been driven off, the women and children were rounded up and the children systematically killed by impaling the older ones and smashing the smaller infants against trees. Death rates through raids of this kind were high. One estimate is that 7 per cent of the women and 24 per cent of the men died in this way. But why so if there is no pressure on natural resources? Chagnon concluded that a predisposition to warfare is ingrained in the unconsciousness of the system, 'military ideology and the warfare it entails function to preserve the sovereignty of independent villages in a milieu of chronic warfare'.

Chagnon's militaristic interpretation has not gone unchallenged, particularly by critics who argue that violent warfare results from disruptions caused by contacts with the outside world, but there is ample witness evidence of warfare going back generations before the contact period.

One question worth considering is that, given that there is now no pressure on resources, how long would it take without warfare (and its 24 per cent male massacre rate) for the population to reach the holding capacity of the territory? In other words, is the explanation Malthusian after all? But to suggest this is to imply that the Yanomamö elders had the ability to predict the ecological consequences of peace. Perhaps it is better to accept Chagnon's conclusions that the Yanamamö were socially predisposed to warfare.

The second group to be considered here are the Maring of the highlands of New Guinea, studied in the 1960s and 1970s and known particularly through the work of Andrew Vayda. The population, of some 7,000 people in 200 square miles of land, is comparatively dense and is organized in clan clusters of 200–850 people; at the core of each is a group of warriors. Subsistence is based on slash-and-burn cultivation and the raising of tuberous staples, supported by pig husbandry and some hunting and gathering.

The Maring practised two styles of direct conflict – 'small fights' or 'nothing fights' and 'true fights', the first often escalating into the second. The 'small fights', held on a daily basis in border zones, were little more than displays of prowess. The warriors of the two opposing bands, armed with shields and bows and arrows, taunted each other and displayed their bravery by drawing fire. As one observer remarked, they remained 'within range of each other's voices while keeping out of range of each other's deadlier weapons'. When the day's competition was over the two sides would return to their homes.

The 'nothing fights' could easily escalate into 'true fights' in which spears and axes, more effective in close combat and far more deadly, were used. True fights would involve the opposed forces being drawn up facing each other while on the field between, individual combats were fought by warriors from the front ranks. Deaths would occur and new fighters would take the field. Reassembling on a daily basis, this kind of fighting could last for weeks and months, while all the time the women stayed at home to tend the gardens. Heavy rain could stop play and rest days were allowed so long as they were agreed by both sides, but if one side failed to turn up this was regarded as capitulation and routing would ensue.

Routing involved an all-out attack on the enemy's village, the burning of the houses and destruction of the gardens and the slaughter of all the men, women and children who could be caught. Of the twenty-nine conflicts on record before 1960, nineteen ended in a rout. Those who survived would disperse and perhaps join other tribes, or they might eventually return. Peace would often involve the redistribution of land with the displaced population settling well away from the new border, but on some occasions they might return to their original settlements and old borders would be re-established.

Given the high density of the population and the consequent pressure on resources, Vayda concluded that warfare among the Maring is an adaptive mechanism brought into being by population stress. But there are other factors at work. The self-image of the group or aspirations of individual group members might create a *perception* of the territory appropriate to support the community that is in excess of the *actual* ecozone required to provide sustenance for it. Simon Harrison, in his *Mask of War*, has given a more subtle interpretation. He sees warfare and its attendant rituals as a process used to suspend the complex network of social relationships, developed between dispersed groups, in the interests of redefining the political unit and its hierarchy. To him, warfare is a positive force for maintaining group coherence.

The issues are clearly complex, but can be crudely simplified in the question – is warfare simply a response to ecological pressures or to some perception of self or group in relation to others (and to appropriate space), orchestrated by naturally aggressive alpha males? In reality, far from being polarized options, these contentions may be seen as part of a continuum.

It is difficult, I believe, to do other than to accept that humans have evolved, through the processes of natural selection, to be aggressive and competitive on the one hand and gregarious on the other. The varied social systems which have developed around the world strive to contain that aggression, so that it is not destructive of the group, and channel it to ensure that the group maintains its integrity and identity. The means of control is structured warfare, externally focused.

The strength of the anthropological data is that it allows us access to first-hand observations of human behaviour throughout the world during the last 250 years or so. It is, however, open to the criticism that we are observing non-state societies only at the stage when they are beginning to interact, usually in competitive mode, with complex societies. In other words, we have access only to special-case scenarios. To provide a real time depth to the enquiry we must now turn to the archaeological evidence.

To begin near the beginning

The archaeological database is massive and grows in volume and sophistication daily. In this brief survey we will restrict ourselves to Europe and its approaches over the last 40,000 years or so.

Hunter-gatherer societies

For most of human history, up to roughly 10,000–7,000 years ago, Europe and the Near East were comparatively sparsely inhabited by hunter-gatherer communities maintaining a high degree of mobility often ranging over considerable territories. Human remains are by no means common, but among the available collections the bones show a surprisingly high incidence of trauma of the kind likely to have resulted from aggression. In the cave of Shanidar in northern Iraq, a small group of Neanderthals were buried about 40,000 years ago. One had died of blows to the head, another was found with a stone blade between the ribs. At Grimaldi, in Italy, the body of a child of the Aurignacian period (about 30,000 years ago) was found with a stone projectile point embedded in its spinal column, while human skulls of the same period from France had been scalped.

These isolated examples are anecdotal evidence of violence. Rather more systematic evidence comes from the Upper Palaeolithic cemetery of Gebel Sahaba in Egyptian Nubia. Of the fifty-nine men, women and children buried there, over 40 per cent had projectile points embedded in, or closely associated with, their skeletons. Several of the adults had multiple wounds and it is estimated that at least half of those buried in the cemetery had died as the result of violence.

The Mesolithic population of Europe was no more fortunate. A survey of the known burials shows that, in Denmark, 44 per cent of the skulls had traumatic injuries (compared with 20 per cent in Sweden and France). At Skateholm, in Sweden – a large hunter-gatherer settlement with associated burials – four individuals had depressed fractures in the skull, which could have been fatal, while two had been hit by flint arrows found among their bones, one in the stomach and one in the chest.

Finally, to conclude this gruesome sample, we should mention the cave site of Ofnet in Bavaria where, in 1908, two 'nests' of skulls were found packed together in pits, thirty-one in one and six in the other. The heads had been deliberately severed from the bodies and buried some time about 6400 BC. Men, women and children are represented and most showed vivid evidence of having been bludgeoned to death.

The interpretation of this evidence is, of course, open to debate. It could be argued that the Ofnet find was the result of ritual sacrifice and that the other observations reflected an accident-prone lifestyle, but this sounds like special pleading. It is simpler to take the evidence on its face value as indicative of widespread warfare among the early hunter-gatherer groups. The frontal injuries suffered by the Skateholm population and the massacre of men, women and children at Ofnet are entirely consistent with the confrontational battles and the routs of the Yanamamö and Maring.

Early farmers

From about 6000 BC more settled economic regimes involving domestication and cereal cultivation spread throughout Europe. During this time, violence if anything seems to have intensified, quite probably because of population pressure and increasing competition for resources. One Neolithic village at Talheim in the Neckar valley, belonging to the LBK culture, had evidently suffered an attack in which at least thirty-four people died. The bodies were thrown, jumbled, into a pit. There were eleven men, seven women and sixteen children. Twenty had suffered blows to the head and in two cases flint arrowheads were embedded in the skeletons. This is surely the result of a rout worthy of the Maring. Nearer at home the defences of a Neolithic camp on Crickley Hill in Gloucestershire seem to have borne the brunt of an archery attack, as witnessed by a large number of arrowheads found on the face of the rampart. Here, however, there is no evidence to suggest that the inhabitants succumbed to the onslaught.

The data we have from across Europe for the period up to about the third millennium are consistent with endemic warfare, of the comparatively simple type seen among contemporary pre-state societies, involving border skirmishes, raids and routs. It was opportunistic, probably badly led, and would have been focused on group definition, the display of dominance, and the acquisition of land, women and other commodities. It was essentially small scale, using the tools of the chase and everyday life – bows and arrows, spears and axes – rather than specially designed anti-human weapons.

The rise of heroes

In the second millennium BC change is apparent throughout most of Europe: warfare is no longer purely opportunistic but becomes increasingly bureaucratic. It is now possible to distinguish two separate traditions, one focused on the east Mediterranean where early states are beginning to develop, the other on 'barbarian' Europe dominated now by the growth of strong chiefdoms.

The range of evidence for warfare now becomes immeasurably richer. Perhaps most dramatic is a huge increase in the number of weapons, and particularly of weapon-sets designed specifically for hand-to-hand fighting. While arrows and spears continue much as before, the panoply now includes a range of elaborate battle axes and, after a brief experiment with halberds and rapiers, the beginning of what is to become a long tradition of slashing swords. Close engagement fighting takes with it the need to protect the body with shields, helmets and other body armour. Much of what survives in the archaeological record is parade armour made in bronze of the kind designed to aggrandize the wearer. Flashy trappings of this kind would not have given much protection in battle unless backed by more resilient materials such as leather. For day-to-day use, tough leather gear would probably have been widely adopted by those unable to aspire to elite metal armour.

The Bronze Age warrior, resplendent in all his finery, is charmingly illustrated by contemporary figurines from Scandinavia and in particular from the Nuragic culture of Sardinia. In south-western Iberia, at about this time, it was customary to indicate warrior prowess by depicting the attributes of status (swords, shields, vehicles, etc.) on stelae which were probably used to commemorate individual warriors.

The elaboration of the weapon-sets leaves little doubt that we are dealing with a type of warfare in which individual heroes dominate the stage, men who engaged in single combat in full view of their

FIGURE 4.1: Portrayal of violent conflict in the prehistoric rock carvings of Val Camonica, northern Italy (© Andrea Arca)

supporting forces in much the same way as the 'true fights' of the Maring centred around duelling warriors in single combat. Indeed, actual conflicts of this kind are depicted on contemporary rock carvings found as far apart as Val Camonica in northern Italy (Figure 4.1) and Bohuslän in Sweden. That said, the sheer volume of weapons found throughout Europe leaves little doubt that warfare was widespread both spatially and also socially: it was not restricted solely to the elite.

By the end of the second millennium, raids and all-out attacks on settlements had become far more commonplace. This much is implied by the rapid growth in the number of defended settlements found throughout Europe ranging from the hillforts of northern Europe, built with ramparts of earth and timber, to the stone-built towers and stone-built curtain walls with projecting bastions which are to be widely found on Sardinia.

No doubt in all this – the dominating defensive architecture and the elaborate weapon-sets – there is a degree of display involving ceremonial and parade. These are the trappings of simulated aggression, but behind it all lies a society in which prestige and status in daily life are measured by bravery and prowess in battle and in which the elite are distinguished by the size of the group of followers they could attract. Some sense of the reality of this world can be gained by the scenes of Bronze Age warfare so vividly presented in Homer's *Iliad*, from the power of individual leaders, and their tetchy reactions to slights and insults, to the massed onslaught on Troy. Pervading it all is the central importance of honour and the display of personal prowess in single combat. The engagement of the warrior-hero Patroclus and 'bronze-clad' Hector before the walls of Troy can be taken to be symbolic of Bronze Age heroic society. The picture which Homer gives of Patroclus preparing for the fatal combat is timeless:

> Patroclus put on the shimmering bronze. He began by tying round his legs the splendid greaves, which were fitted with silver clips for the ankles. Next he put on his breast Achilles' beautiful cuirass, scintillating like the stars. Over his shoulders he slung the bronze sword, with its silver-studded hilt, and then the great thick shield. On his sturdy head he set the well-made helmet. It had a horsehair crest, and the plume nodded grimly from on top. Last, he picked up two powerful spears which suited his grip.
>
> Homer, *The Iliad*, trans. E. V. Rieu, Harmondsworth: Penguin, 1950, 295–6

The style of warfare depicted in the Homeric epic owes much to memories of the Mycenaean Late Bronze Age – a time when the Aegean states were developing more complex forms of warfare. This involved the adoption of specialized equipment, like the light two-wheeled vehicle for moving warriors efficiently about the field of battle, and the creation of organized troop detachments prefiguring the hoplites of the later Greek city states. The bureaucracy now involved in this institutionalized style of warfare is most clearly shown in the Linear B tablets from Pylos from which we gather that supplies of military equipment were now being centrally organized and complex hierarchies were now required to control military operations. In all this we are witnessing the beginning of professional warfare.

The northern barbarians: Celts, Germans and others

In the first millennium BC the divide between the civilized Mediterranean, with its states and nascent empires, and the rest of barbarian Europe, still tribal and dominated by chiefdoms, becomes more marked. The northern barbarians, characterized as Celts, Germans and Scythians by Classical writers, now come into much sharper focus through the writings of the Greeks and Romans who encountered them. There are, of course, difficulties in dealing with these texts, not least because of the desire of Classical authors to depict their principal opponents – in particular the Celts – as 'other', that is different from 'us', but still a worthy foe, ferocious if rather naïve. Thus in battle they are emotional and unstable and often fight naked (we are controlled and rational). They have long shaggy hair, made stiff by rubbing in lime, and long drooping moustaches (we have no tonsorial eccentricities) and they drink their wine undiluted (unlike we self-respecting Mediterraneans). Yet they are a force to be reckoned with – a force we can be proud to overcome. The Classical 'topos', then, provides something of a caricature, and yet it is possible from the texts to build up an intricate picture of barbarian warfare in the latter part of the first millennium BC.

The central element in traditional Celtic warfare was single combat between opposing heroes. Such a system was carefully recorded by the ethnographer Poseidonius in the second century BC:

> When the armies are drawn up in battle array, they [the chiefs] are wont to advance before the battle line and to challenge the bravest of their opponents to single combat, at the same time brandishing before them their arms so as to terrify their foe. When someone accepts their challenge to battle, they loudly recite the deeds of valour of their ancestors and proclaim their own valorous quality, at the same time abusing and making little of their opponent and generally attempting to rob him beforehand of his fighting spirit.
>
> Diodorus, *Hist.* 5.29

After this preliminary, the combats commence in full view of the opposing sides. Sometimes, after a day of contests, the forces would disperse, but on other occasions the heat of the occasion would ignite and a mêlée of slaughter would ensue.

A rather different aspect of warfare was the raid. Proposals to mount raids took place at public assemblies or sometimes at feasts.

> When a chieftain in council says that he will lead a foray and asks for followers, those who approve of the man and his project rise to

support him and to promise their help, and are applauded by the
assembly. Those who fail to keep their promises are looked on as
deserters and traitors and are henceforth trusted in nothing.

Caesar, De Bello Gallico, VI, 23

In Celtic society the size of a man's group of followers demonstrated
his status and the success of his enterprise would ensure a greater
number would join him the next time he proposed to lead a raid. With
status directly linked to size of entourage, raiding became a socially
competitive enterprise, each aspiring leader seeking to outdo his rival.
In this way, raiding became socially embedded with cycles of raiding
tending to escalate. It is against this background that we might explain
the 'Celtic migrations' which saw sectors of the population moving
permanently away from their homes to set up new enclaves. From such
bases, in the Po valley, the Middle Danube region and central Anatolia,
raiding cycles continued on an annual basis with the cities and sanc-
tuaries of the Roman, Greek and Hellenistic worlds becoming the
targets. In these situations the warrior bands would have become
entirely removed from productive activities. The Roman historian
Tacitus, writing of the Germans at the end of the first century AD,
sums up the situation there. The German, he says,

is not so easily prevailed upon to plough the land and wait patiently for
the next harvest as to challenge a foe and earn wounds for his reward.
He thinks it tame and spiritless to accumulate slowly by the sweat of
his brow what can be got quickly by the loss of a little blood. . . . The
boldest and most warlike men have no regular employment, the care of
the house, home and fields being left to women, old men and
weaklings of the family.

Tacitus, Germania, 14

The status of the tribe also required the assumption of an active
aggressive stance towards neighbours:

What the states greatly admire is that the lands surrounding them
should be devastated and left solitary to the furthest extent. This, in
their eyes, is the true proof of valour, that their neighbours should be
expelled and driven to flee from their fields, and that none should dare
to take up his stand near them.

Caesar, De Bello Gallico, VI, 23, 1

Reading the accounts of Classical writers about their northern neighbours leaves little doubt that warfare was institutionalized within the
social system. Of this the Greek geographer Strabo was clear. Writing
of the Celts he concluded:

> The whole race . . . is madly fond of war, high spirited and quick to
> battle, but otherwise straightforward and not of evil character. And so
> when they are stirred up they assemble in their bands for battle quite
> openly and without forethought . . .
>
> Strabo, *Geog.* 4.4.2

Rome: the state and the development of the standing army

The history of the Mediterranean, in the last few centuries BC, was
dominated by the growth of Rome and the extension of her power first
throughout and then beyond the Italian peninsula. The social system
of early Republican Rome had many similarities to that of the Celtic
world, not least the fact that in both societies high social prestige and
status could be obtained only through military achievement. In the
second century BC, all office-holders in Rome were expected to be
competent military leaders. It was normal for any man intent on a
public career to enter the army and to serve for five to ten years. During
this time he would expect to command a force which, if effectively
managed, would lead to foreign conquests, enhanced reputation, booty
and – so long as he had won a significant battle in which at least 5,000
enemy were killed – he would be awarded the accolade of a triumph in
Rome. This is not very different from the Celtic war leader who built
his reputation on a succession of profitable raids. The difference is
that in Rome the system became firmly embedded in the developing
state system. As more men competed for status, so more theatres of
war had to be opened up.

The effect of all this was that during the last two centuries of the
Republic the state was almost continuously at war. As the ancient historian Keith Hopkins has written, 'The Roman elite was permeated
with pride in its military achievements' and triumphal monuments
were everywhere to be seen in the city. In the period 252–53 BC no less
than seventy military triumphs were staged predicated on the death of
a minimum of 350,000 enemy!

It is against this background of state-propagated aggression that Julius Caesar could operate with impunity in Gaul, Germany and south-eastern Britain between 58 and 51 BC. A few statistics taken from his account of his battles in Gaul are revealing:

58 BC Of 300,000 Helvetii who attempted to migrate but were beaten back by Rome only 100,000 returned home.

57 BC After bitter fighting against the Nervii only 500 of the population of 60,000 survived.

57 BC Against the Aduatuci 4,000 killed and 53,000 sold into slavery.

56 BC Against the Veneti the elders were put to death, the rest sold as slaves.

52 BC After the siege of Cenabum all but 800 of the 40,000 population killed.

51 BC At Uxellodunum all those who had carried arms against Rome had their right hands cut off.

According to Plutarch, in less than ten years, 'Caesar had taken by storm 800 towns, subdued 300 states. Of the 3 million who made up the gross sum of those engaged he had killed one million and captured a second.' Even allowing for some exaggeration this was an impressive ten years' work. Given the estimated population of Gaul of about 6 million Caesar's activities would have been massively disruptive.

The effects of state-sponsored aggression on Rome itself were not inconsiderable. In the early second century BC out of a total male population of about a million it is estimated that 100,000 Romans and allies had been killed. Keith Hopkins' calculations of the size of the army are staggering. Over the last two centuries of the millennium, the army would have comprised on average 13 per cent of the citizen population, that is equivalent to enlisting 84 per cent of all seventeen-year-olds for five years or 44 per cent for ten years. In terms of actual size, adding in the contribution of the Italian allies, there would have been about 130,000 men on active service at any one time. This is about the size of the French army in the mid-seventeenth century, then the largest in Europe (the only difference being that the base population of France was three times greater than Rome).

Late Republican Rome was then a warrior state driven by the competing ambitions of a militaristic elite, the whole sustained by the wholesale recruitment of its peasant population. The long-term effects of all this were massive: the peasants became impoverished and moved to the towns to become the urban poor while rural labour was increasingly supplied by slaves (a product of war). This change, combined with the influx of booty and its investment in land, led to the creation of great estates and so the structure of the Empire came into being sustained by a fully professional standing army permanently maintaining the distant frontiers and engaged in continuous warfare safely distanced from the Mediterranean heartland.

Epilogue

In the comparatively peaceful core of the Empire the cities harboured large populations, many of whom were poor, their aggressive instincts now catered for by spectacles put on to entertain them in circuses and amphitheatres.

In AD 59 the citizens of Pompeii were enjoying a gladiatorial display, together with visitors from the neighbouring town of Nuceria, when, in the rivalry and excitement of the contest, taunts were exchanged and soon missiles began to be thrown. The situation rapidly got out of control and in the chaos which followed many of the visitors were killed and mutilated. The whole event was depicted on a wall painting soon afterwards. So serious was the riot that the emperor Nero set up a senatorial commission which resulted in a ten-year ban on all games at Pompeii and the exile of those responsible for organizing the event.

The relevance of this anecdote will be evident. There is not really much difference between the contest of the gladiators in the amphitheatre of Pompeii, urged on by the two supporters' groups, and heroes locked in single combat watched by the rival forces, whether they be Greeks and Trojans before the walls of Troy, Celtic chieftains in central Europe or Maring in the highlands of New Guinea – or, for that matter, football teams playing at Millwall. In each case, the aggression of the opponents is contained by agreed rules and the contest is played out by the appointed champions in full view of all. Often, at the end of the day, everyone packed up and went home, but on some occasions passions would rise, the constraints would give and the rival fans would invade the pitch.

If our all-too-brief look at the roots and development of warfare has taught us anything, it is how timeless and transglobal are the confrontations forced on us by the deeply embedded instincts that we aggressive humans have acquired through natural selection. Only a delicate tissue of social constraint, carefully nurtured, prevents our lives from becoming 'nasty, brutish and short'.

FURTHER READING

N. Chagnon, *Yanomamö: The Fierce People*, 3rd edn. New York: Holt, Rinehart and Winston, 1983.

A. Flew (ed.), *Malthus: An Essay on the Principle of Population*. Harmondsworth: Penguin Books, 1970.

M. Fried, M. Harris and R. Murphy (eds.), *The Anthropology of Armed Conflict and Aggression*. New York: The Natural History Press, 1967.

J. Haas (ed.), *The Anthropology of War*. Cambridge: Cambridge University Press, 1990.

S. Harrison, *The Mask of War: Violence, Ritual and the Self In Melanesia*. Manchester: Manchester University Press, 1993.

L. H. Keeley, *War Before Civilization*. Oxford: Oxford University Press, 1996.

S. A. LeBlanc, *Constant Battles*. New York: St Martin's Press, 2003.

D. L. Martin and D. W. Frayer (eds.), *Troubled Times: Violence and Warfare in the Past*. London: Gordon and Breach, 1997.

R. Osgood, S. Monks and J. Toms, *Bronze Age Warfare*. Stroud: Sutton Publishing, 2000.

A. Vayda, *War in Ecological Perspective*. New York: Plenum, 1976.

5 Conflict in the Middle East

LISA ANDERSON
School of International and Public Affairs, Columbia University

> Middle East – also Mideast. An area comprising the countries of
> southwest Asia and northeast Africa. In the 20th century the region
> has been the continuing scene of political and economic turmoil.
>
> American Heritage Dictionary[1]

This is the definition of the Middle East given by a dictionary widely
used in the United States. For the authors of this dictionary, as for
many of its users, conflict is not merely a descriptive attribute of what
we call the Middle East, but part of its very definition.

How did this come to be? Can it be that a geographical region is actually defined by the quality of its politics? Would there be no Middle East
were there no turmoil there? Would we call this region South-west Asia
perhaps? What *is* in a name? And what, in the final analysis, is all this
turmoil?

I will suggest that this place we call the Middle East – which now
seems to be both East of nowhere and yet in the Middle of everything
– has been, since the very term arose, as much an imaginative invention as an actual geographical place. As Ghassan Salame, one of the
region's leading political scientists, put it:

> the Middle East is a zone whose boundaries are practically impossible
> to demarcate; thus it is impossible to locate the players to be taken
> into account when identifying equations of security or causes of
> insecurity. Should, for example, the Maghreb be included in this
> equation? Does Iran pose a military threat to Israel or Egypt? Is the
> Horn of Africa a real source of concern for Cairo? Should we take
> account of Turkey in the regional balance of power? Has the collapse

---

[1] *The American Heritage Dictionary of the English Language*, 4th edn. New York:
Houghton Mifflin Company, 2000.

of the Soviet empire really pushed Central Asia into the Middle
Eastern strategic equation? Such questions provoke highly varied
responses from one leader to another, from one analyst to another,
cloaking the zone itself in a geo-epistemological fog which renders any
attempt to define the theatre and cast of players nearly impossible.[2]

Indeed, the boundaries of this region *are* impossible to demarcate,
precisely because our imaginations have endowed it with various and
vastly different meanings and therefore with different shapes and
sizes. As an imaginary place, the Middle East has been assigned
myriad hopes and dreams, aspirations and ambitions. The geographical region, criss-crossed with political and social boundaries, both
arbitrary and significant, has been shaped by our intangible frames of
minds, and these, it seems, have been in turmoil.

What is the Middle East?

Before we examine the nature of the unease and anxiety that has contributed to this conflict, let us look at where the designation 'Middle
East' originated and what has happened in the region it signifies over
the last century or so. The term 'Middle East' is a term of art, utilized
in virtually all languages and by almost all international organizations,
to refer to the region between Libya and Iran, sometimes including
North Africa, sometimes Afghanistan, almost always Turkey and occasionally Sudan. It is a part, but by no means the largest part, of the
Muslim world, it houses a large number, but by no means the majority, of the world's Jews, and it includes a small but substantial fraction
of the world's Christians. For these, the world's major monotheistic
religions, it is also the site of their principal holy places. This region
also includes most of the world's Arabs, but also many non-Arabs –
Turks and Persians, Kurds and Berbers, Druze, Azeris, Baha'is,
Circassians and many, many more. 'Middle East' is a term that privileges no one in the region, which may be part of its utility, but it does
evoke the continuing shadow of the area's European imperial past.

From the perspective of Europe, this region has always been part of
the 'East', where the sun rises. Indeed, early Renaissance Italian

---

[2] Ghassan Salame, 'The Middle East: elusive security, indefinable region', *Security
Dialogue* **25:1**, 1994, 19.

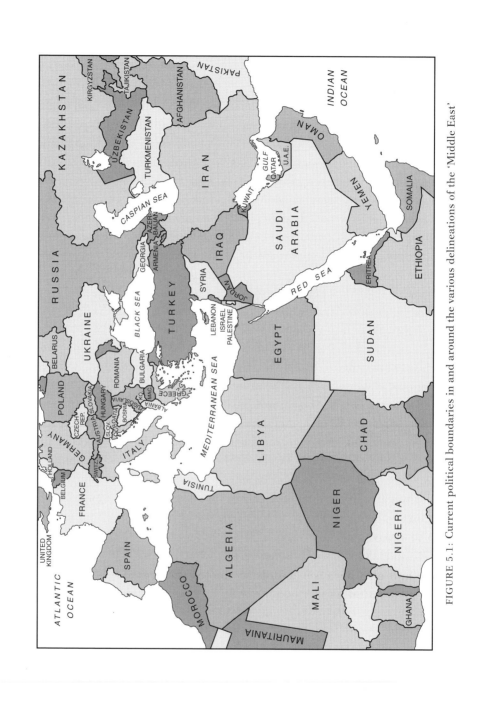

FIGURE 5.1: Current political boundaries in and around the various delineations of the 'Middle East'

traders referred to the eastern Mediterranean as the 'Levant' and that term and its derivatives – 'Levantine', for example – are still occasionally heard. By the nineteenth century 'the Orient' was a common usage, more frequently referring to the distant lands of East Asia, but anything exotic to European eyes – even Algeria, which is, of course, due south of Europe – could be encompassed in the mysterious 'Orient'. In European political and strategic circles, however, the 'Near East' was applied to the territories of the Ottoman Empire, in contrast with the Far East of Japan, China and Indo-China. The vast expanse in between was 'India'.

Perhaps not surprisingly, it was an American who coined the term 'Middle East'. In 1902, Captain Alfred T. Mahan used the term for the region between India and the Near East – essentially what we now know as the Persian – or, sometimes, Arabian – Gulf. After World War I, and with the demise of the Ottoman foundation of the 'Near East', the British increasingly applied the term 'Middle East' to the territories from the central Mediterranean to the Indian subcontinent. During World War II, the headquarters of the British Middle East Command was located in Cairo. By the 1950s, after a bit of temporizing about a place called the 'Near and Middle East', nearly everyone had followed suit, and the Middle East took on its modern, contested meanings.

In the 1970s and 1980s, there was a brief effort in academic circles to throw off the residual Eurocentrism of the term in favour of 'South-west Asia' or even SWANA – an acronym for South-west Asia and North Africa – and for reasons both obvious and complex, it failed. A current textbook concludes its discussion of the term 'Middle East' with a revealing, if somewhat apologetic, justification: 'The toponym is used in this book, both because it is well accepted in international usage and also because 'middle' connotes the region's central location, its function as a tricontinental hub, and its role as a strategic bridge'.[3]

3 Colbert C. Held, *Middle Eastern Patterns: Places, Peoples, and Politics.* Boulder, CO: Westview Press, 2000, 8.

How much conflict is in the Middle East?

A strategic bridge it may be, but there is considerable irony in our association of the Middle East with conflict, since the twentieth century has been far bloodier elsewhere in the world. Indeed, in one listing of the thirty deadliest armed conflicts of the twentieth century, the Middle East – broadly defined – appears at only at numbers 15 (the Soviet invasion of Afghanistan); 18 (the Iran–Iraq war of 1980–88); and 23 (the Algerian war of independence), accounting for 3,175,000 deaths in a total of about 175 million. If we disaggregate the Armenian massacres from World War I and add a few of the smaller conflicts – the Italian conquest of Libya, for example – we might be able to add perhaps 1.5 million to the Middle Eastern ledger.

These 5 million souls represent a significant number of deaths but, as a proportion of total population, both Africa and the area encompassed by the unlamented Union of Soviet Socialist Republics saw far greater devastation in the twentieth century. Conflicts in Congo at the beginning and the end of the century contributed to the death of perhaps as much as a quarter of the total population. Poland during the Second World War, Cambodia during the 1970s and the countries of the Soviet Union under Stalin each lost nearly as large a proportion of their people. In Mexico during the Mexican revolution of 1910 to 1920, a million people – more than 10 per cent of the population – perished. By contrast, the Arab–Israel wars – 1948, the 1956 Suez war, the wars of 1967 and 1973, the first and second *intifada*s and general civil strife during Israel's first fifty years – which loom far larger in our collective memory than, say, the Mexican Revolution, contributed to about 70,000 deaths.[4]

Of course, conflict is not measured merely by absolute numbers of battlefield casualties. Terrorism, that slow, corrosive but always media-genic tactic, may also be a measure of political turmoil. Constant news of what we now call improvised exploding devices, of hijackings and kidnappings have certainly heightened the image of the Middle East as conflict-ridden. Yet even as early as the mid-1980s,

---

[4] See Matthew White, 'Death tolls for man-made megadeaths of the twentieth century', 2004. http://users.erols.com/mwhite28/atrox.htm; http://users.erols. com/mwhite28/warstats.htm.

when one of the Reagan Administration's policy architects reported that, 'in the Middle East . . . terrorism continued at much higher levels of frequency and lethality than in any other region' he also noted, in a puzzling aside, that, 'more incidents involving American persons or property took place [in Latin America] than in any other region'.[5] Walter Laqueur, an authority on terrorism, points out that 'terrorist suicide attacks' are hardly unique to the Middle East, reporting that when the Japanese authorities asked for kamikaze candidates during the last year of World War II, 'many thousands volunteered and some 4,600 were killed'.[6]

According to the US Committee on Refugees, the Middle East hosted the largest number of refugees in the world at the end of the twentieth century, nearly 6 million. Of those about 3.8 million were Palestinian; Iran was home to another 1.8 million refugees from Iraq, Afghanistan and elsewhere. There are also substantial numbers of internally displaced persons in the Middle East – from the Iraqi Kurdish and Shi'i populations forced to move in the 1990s to the thousands of Algerians placed in camps in the south after aborted elections in 1991. Yet even these statistics are something of an artefact of regional boundaries; in the late 1990s, Africa hosted fewer than 3.1 million people registered as 'refugees' but more than 10 million internally displaced persons.[7]

We need not spend more time on gruesome statistical competitions. The point is clear: our sense of the region as characterized by conflict is at least partly a reflection of what we expect to see. The Middle East is a region in turmoil because we have defined it that way, assigning the lives lost there particularly high values, taking special note of the plight of the survivors there, even extending its borders to include new conflicts. As it descended into war, for example, even Afghanistan, once safely tucked away in Asia, seems to have migrated to the Middle East.

[5] Robert Oakley, 'International terrorism', *Foreign Affairs*, America and the World, 1986, http://fullaccess.foreignaffairs.org/19870201faessay7834-p0/robert-oakley/international-terrorism.html.

[6] Walter Laqueur, 'Reflections on terrorism', *Foreign Affairs*, Fall 1986, http://fullaccess.foreignaffairs.org/19860901faessay7808-p0/walter-laqueur/reflections-on-terrorism.html.

[7] US Committee for Refugees, http://www.refugees.org/; Dan Smith, *The Atlas of War and Peace*. London: Earthscan Publications, 2003, 66.

What is the Conflict in the Middle East about?

There are many who see in conflicts which we have assigned to the Middle East an underlying confrontation so deep as to be what they call 'civilizational'. Bernard Lewis, the great Orientalist, has argued, for example, that conflict between the West and Islam is inevitable. What he calls 'Muslim rage' is, he tells us, 'a rejection of Western civilization as such, not only what it does but what it is, and the principles and values that it practices and professes. . . . This is no less than a clash of civilizations – the perhaps irrational but surely historic reaction of an ancient rival against our Judeo-Christian heritage, our secular present, and the world-wide expansion of both'.[8]

Samuel P. Huntington, the well-known American political scientist, seized upon Lewis' observation and extended the argument. 'Conflict along the fault line between Western and Islamic civilizations has been going on for 1,300 years', he contends, and today 'in Eurasia the great historic faultlines between civilizations are once more aflame' As he famously put it, 'Islam has bloody borders'.[9]

But if we examine a few of these borders, we notice on a map of the Middle East today approximately two dozen sovereign states. By and large, these states are members of the United Nations. Their governments issue passports and guard internationally recognized frontiers. They are presumed to maintain law and order, regulate national markets, and provide for social welfare. The vast majority of these states divide Muslims from each other rather than from representatives of Western or Judeo-Christian civilization. More importantly, no more than one or two of these states were sovereign countries eighty or ninety years ago and most did not exist at all.

In fact, the turmoil in the Middle East is not between civilizations but within them, within societies and traditions that are struggling to define meaning and membership. To be sure, some of the challenge that struggle represents reflects the remarkable influence exercised by Western powers around the world over the last several centuries. It

---

[8] Bernard Lewis, 'The roots of Muslim rage', *The Atlantic Monthly*, September 1990, available on-line at *Policy*, http://www.cis.org.au/policy/summer01–02/PolicySummer01_3.html.

[9] Samuel P. Huntington, 'The clash of civilizations?' *Foreign Affairs*, Summer 1993.

was, after all, Western powers that drew most of the lines on the modern map. But the Middle East reveals – indeed, seems to evoke – conflict within the West just as it does within the traditions of the region itself.

The fortunes of the modern state in the Middle East

To examine what is at issue behind this familiar map of states, we may begin in the nineteenth century, when very different mechanisms for maintaining law and order and fostering prosperity prevailed in the region. There were no states in the conventional sense, although European kings and prime ministers often trucked with local sultans and shaykhs as if they played analogous roles. The modern European state had arisen in Europe only in the sixteenth century, triumphing over alternative arrangements – kinship-based aristocracies, feudal arrangements, trade networks, even the Church itself. Although that victory was implicitly acknowledged in the Peace of Westphalia (1648), which is commonly cited as the origins of today's interstate system, elsewhere in the world, including the Middle East, such alternative devices for organizing human society continued to flourish.

By the twentieth century, however, and courtesy of European imperialism, most of the globe was claimed for a European-style interstate system. The League of Nations, established after World War I, recognized only two kinds of political unit. Gone were religious empires, trading cities, tribal federations. There were only independent states, like those in Europe, and, as the League Covenant put it, territories 'which are inhabited by peoples not yet able to stand by themselves under the strenuous conditions of the modern world'. On this basis the European powers dissolved the Ottoman Empire, which had controlled much of the territory known as the Middle East for nearly four centuries, and establish in its stead 'independent nations' that were to 'be provisionally recognized subject to the rendering of administrative advice and assistance by a Mandatory until such time as they are able to stand alone'.[10]

---

[10] Article 22, *Covenant of the League of Nations*, 28 June 1919.

89

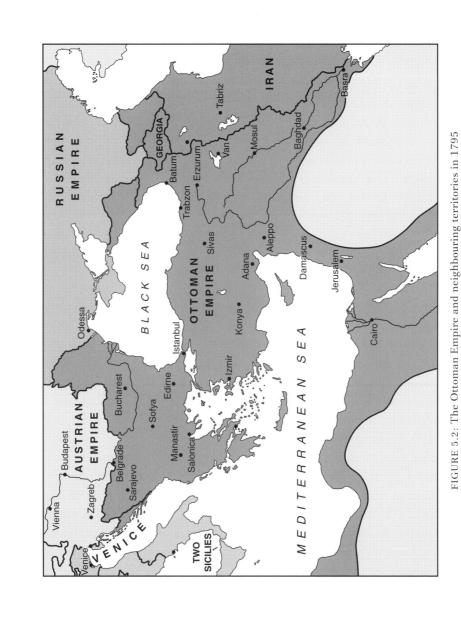

FIGURE 5.2: The Ottoman Empire and neighbouring territories in 1795

The story of the League Mandates in the Middle East is too well known to require rehearsal here. Territorial units – Syria, Palestine, Iraq, for example – were carved from the Empire with little regard for the political identities or aspirations of local populations, except for communities already familiar to the European 'advanced nations' to whom the Mandates were entrusted. The French did extract Lebanon from its Syrian Mandate in deference to the wishes of the Maronite Christians and the British divided the Palestine Mandate into two parts – Palestine and TransJordan – in order to fulfil the promise of the World War I Balfour Declaration in support for a Jewish homeland. For the rest, states were drawn up in the expectation that territorial identity – Iraqi, Syrian, Jordanian – would trump other 'obsolete' loyalties.

This was to mean that, whatever doubts the peoples of the region might have harboured about the polities, aspirations to be rid of European domination were couched in terms of independence for these political units. Only states, understood as these territorial units, could hope to join the 'advanced nations' and so independence movements organized themselves within the new state boundaries, from Palestine to Morocco.

By the 1940s, for example, the ambiguity of the formulation of the 'Jewish homeland' seemed self-defeating. As the World Zionist Organization put it in the 1942 Biltmore Program, since 'the new world order that will follow victory cannot be established on foundations of peace, justice and equality, unless the problem of Jewish homelessness is finally solved', it was now essential that 'Palestine be established as a Jewish Commonwealth integrated in the structure of the new democratic world', that is to say, an independent state.[11] Similarly, claims for independence in, say, Morocco played down the complex sources of the monarch's legitimacy as commander of the faithful. In 1944, the Moroccan nationalist movement appealed to 'the liberal and comprehensive spirit that animates all Frenchmen of the Resistance' to convey their 'demand [for] the independence of

---

[11] 'The postwar Zionist (Biltmore) program', cited in J. C. Hurewitz (ed.), *The Middle East and North Africa in World Politics: A Documentary Record*, vol. II. New Haven: Yale University Press, 1975, 597.

Morocco in its territorial integrity under the crown of His Majesty Sidi Muhammad bin Yusuf, may God glorify him'.[12]

This strategy worked; the European powers evacuated the states they had erected and left them in the care of local authorities they hoped could 'stand up to the strenuous conditions of the modern world'. For the first several decades after independence, the template of the *state system*, as opposed to the specific states themselves, remained powerful. Arab nationalists debated the merits of the particular states they had been bequeathed, experimenting with efforts like the merger of Syria and Egypt into the United Arab Republic, but they did not confront the notion of the state itself. Indeed, most of the political turmoil was associated with efforts to consolidate power within the new states.

For the most part, though the states seemed fairly hardy, the influence of European models of government did not survive the end of European rule. Liberal democratic institutions were quickly distorted or destroyed by regimes bent on consolidating power. Turkey, which did not fall under European control, became a republic in the 1920s and a democracy by the 1950s, though one in which the military took an unusually public proprietary interest. Among the Ottoman provinces that had been refashioned as League of Nations mandates, Lebanon's democratic institutions, designed to privilege Christian constituencies, splintered under pressure; Israel's non-Jewish populations remained under military rule for more than a decade and faced legal disabilities thereafter; the monarchies of Jordan and Iraq quickly dispensed with their constitutional restraints.

Other European possessions, including Morocco, Tunisia, Libya and Egypt, also gained independence in the 1940s and 1950s as constitutional monarchies, but only Morocco's king survived the upheavals of the next several decades, and even he shed most of his constitutional fetters. Algeria's war of independence did not produce even lip-service to democracy but rather a socialist regime that was quickly overtaken by military rule. When the British finally withdrew from

<hr>

[12] 'Manifesto of the Istiqlal Party, 11–Janaury 1942', cited in Hurewitz (ed.), *The Middle East and North Africa in World Politics*, 690.

'east of Suez' in the early 1970s, there was little effort to create democratic government; most of the small states of the Persian Gulf were
content with appointed consultative councils. Iran and Saudi Arabia,
both of which had remained formally independent throughout the
twentieth century, did so as absolute monarchies; Iran toyed with an
elected parliament but the Shah's government was overthrown, parliament and all, in 1979.

The frailty of democracy was, as it turned out, a signal and symptom
of profound unease with the polities bequeathed the region at independence. As Dankwart Rustow argued, there is only one important condition for democracy – 'that the vast majority of citizens in a
democracy-to-be must have no doubt or mental reservations as to
which political community they belong to'[13] – and that condition was
absent. Political conflict in the region transcended, and could not be
contained in, routine electoral competition. It was not only in Israel
that, in the words of Baruch Kimmerling, 'faced with political problems, [the rulers] saw only existential threats'.[14]

In the 1970s, the cycles of coups and countercoups, unions and
secessions, attacks and counterattacks seemed to abate, as most of the
regimes dug in for the long haul. Israel's control over the territories
occupied in the 1967 war hardened after the 1973 war, and governments throughout the Arab world consolidated their grasp on power.
By the end of the century, however, the costs of the distribution of
power established by this state system were apparent. The region's
economic growth rate for the last quarter of the twentieth century had
been negative and vast numbers of people in the region felt disenfranchised and dismayed. As the historian Albert Hourani had warned
several decades earlier, the failure of governments to do more than
stay in power left a vacuum:

> In most countries of the Middle East, although in some more than
> others, there is growing up a new ruling elite isolated by power and
> education from the mass of the people. The movement of economic

---

[13] Dankwart Rustow, 'Transitions to democracy', reprinted in Lisa Anderson (ed.),
Transitions to Democracy. New York: Columbia University Press, 1999, 26–30.
[14] Baruch Kimmerling, 'Israel's culture of martyrdom', *The Nation*, 10/17 January
2005, 38.

and social development, in its first stage at least, may strengthen the position of this limited class; it may improve their standard of living faster than that of the people, and will certainly give them vast power over the life of society. For their part the mass of the people may find in the policy and nationalist ideas of this class neither a real profit nor an ideal which satisfies the human desire for justice. If this happens, there may rise to the surface a new type of revolutionary spirit. . . . Up to a point, this new spirit may find old channels of expression. The idea of the Muslim community is still alive in the popular mind; the mystical orders are still alive, though not among the educated; and inherent in this popular Islam is an ideal of revolutionary justice and of human virtue.[15]

The acute American analyst Anthony Cordesman put it more bluntly at the end of the century: "a great deal of the explanation of the poor performance of the region lies in the fact that many Middle Eastern states have no enemy greater than their own governments".[16]

The revival of alternatives to the state

Perhaps not surprisingly, disappointment and disillusionment with actual governments begged more fundamental questions about the very purpose of these polities themselves. In fact, political problems *were* becoming existential threats, and as such they reawakened century-old dilemmas. As Hourani suggests, there are alternative identities, different conceptions of political obligation and account-ability, to those embedded in the modern European-style state. Efforts to reconcile these alternatives with the demands of the modern state date back to the French Revolution, and it is instructive to revisit this history.

The Revolution's 'rights of man and the citizen' were to be univer-sal, like the state in which those citizens lived, and like those states, they brooked no exceptions. It was this principle that provoked debate about the necessity of the emancipation of the Jews in France and in voting for emancipation in 1791, Count Stanislas de Clermont-

---

[15] Albert Hourani, *The Emergence of the Modern Middle East*. Berkeley: University of California Press, 1981, 191–2.

[16] Anthony Cordesman, 'Transitions in the Middle East: an address to the 8th US Mideast Policymakers Conference, September 9, 1999'. Http://www.csis.org/ mideast/ reports/ transitions.html.

Tonnerre famously declared 'One must refuse everything to the Jews as a nation but one must grant them everything as individuals; they must become citizens'.[17] The institutions of the Jewish communal life that had been formally sanctioned and protected by the king were relegated to private life as voluntary organizations, allowed to operate insofar as they did not interfere with fulfilment of the obligations of citizenship. The crisis of identity this precipitated for Jews in Europe and eventually around the world was, perhaps ironically, a dilemma that soon faced Muslims as well, for in 1798, Napoleon invaded Egypt proclaiming,

> Peoples of Egypt, you will be told that I have come to destroy your religion; do not believe it! Reply that I have come to restore your rights, to punish the usurpers, and that more than the Mamluks, I respect God, his Prophet, and the Qur'an.[18]

Napoleon's respect for Islam was as profound as Clermont-Tonnerre's was for Judaism: insofar as it did not interfere in the realization of the rights of man and of the citizen, these Frenchmen were content that both Jews and Muslims practise their faith. Of course, for Jews and Muslims alike, the question was not nearly so simple. Both Judaism and Islam impose communal obligations, including observance of religious law. The tension between communal loyalty and individual citizenship has echoed through the next two centuries, reflecting and amplifying the tension between communal identities and statehood.

In Europe, Jewish emancipation proceeded in fits and start throughout the nineteenth century. In Britain, for example, Jews were eligible for election to the Parliament in 1858, allowed to enrol at university in 1870, and extended full civil and political rights in 1890. At the same time that formal legal disabilities were being lifted across the continent, however, social and economic discrimination persisted and, indeed, intensified. The term 'anti-Semitism' was coined in 1879 by a German writer to emphasize, as Mitchell Cohen puts it, 'that his

---

[17] Cited in Mitchell Cohen, *Zion & State*. New York: Columbia University Press, 1992, 52.
[18] 'Napoleon's Proclamation to the Egyptians, 2 July 1798', cited in J. C. Hurewitz (ed.), *The Middle East and North Africa in World Politics: A Documentary Record*, vol. I. New Haven: Yale University Press, 1975, 116.

hostility to Jews was based not on religion but on social and economic justifications, and especially their "foreignness"'[19] – that is to say, the continuing power of communal loyalty and identity.

At nearly the same time, Europeans were touting the very same notion of individual rights in the Ottoman Empire. Like the kings of pre-revolutionary France, the Ottoman sultan embodied both worldly power and religious authority, the latter in his role as Caliph. And like the kings of the *ancien régime*, the sultan granted formal recognition and protection to corporate expressions of recognized religions. As Commander of the Faithful and viceroy of the Prophet on earth, the Sultan was responsible first and foremost for the welfare of the *umma*, the Muslim community. He was also obliged, however, to protect the corporate expressions of certain religious minorities. This was a principle enshrined in Islam; the 'Peoples of the Book' – Christians and Jews – are permitted to follow their own faith. They are not equal to Muslims and are assessed extra taxes that acknowledge their inferiority, but they were also formally recognized and protected as corporate groups. In the Ottoman Empire this was institutionalized in what was known as the *millet* system.

In 1839, and in hope of winning financial assistance from the European powers, the Ottoman sultan inaugurated the reform era known as the Tanzimat with a declaration extending 'guarantees insuring to our subjects perfect security for life, honour and fortune'. Most importantly, the declaration went on, 'these Imperial concessions shall extend to all our subjects, of whatever Religion or sect they may be; they shall enjoy them without exceptions'.[20] In 1856, the Ottoman government reiterated that,

> every distinction or designation tending to make any class whatever of the subjects of my Empire inferior to another class, on account of their Religion, Language, or Race, shall be for ever effaced from the Administrative Protocol. . . . The Taxes are to be levied under the same denomination from all the subjects of my Empire, without distinction of class or Religion . . .[21]

[19] Cohen, *Zion & State*, 44.
[20] Cited in Marvin Gettleman and Stuart Schaar (eds.), *The Middle East and Islamic World Reader*. New York: Grove Press, 2003, 82.
[21] *Ibid.*, 84.

Many of the Ottoman sultan's subjects, Muslim and non-Muslim alike, were appalled. Resistance to this way of defining political community as composed of undifferentiated subjects was apparent throughout the nineteenth century. In 1866, the French rulers of Algeria issued a decree permitting Jews and Muslims to apply for French citizenship; any applicant had to 'declare that he agrees to be governed by the civil and political laws of France'. To many Muslims and Jews, this was interpreted as requiring violation of their religious obligations, since religious law regulates a variety of personal status matters, including marriage and divorce.[22] Over the next fifty years, fewer than 1,400 Algerian Muslims elected French citizenship.

By the end of the nineteenth century, three principal responses to these challenges had emerged. Some European Jews and Muslims in the Ottoman Empire embraced the new conception of communal identity, identified with notions of liberal citizenship and abandoned all but vestigial attachment to religious observance. Others resisted such encroachment on the worldly domain of religious obligation and insisted on the priority of religious identity, vacillating between reactionary postures of refusal and more open gestures towards synthetic arguments. Finally, both Europe and the Ottoman Empire saw the rise of ethnic nationalisms that represented acceptance of fundamentally secular definitions of political community and public life but resisted the individualism of the modern state's definition of citizens.

Within the increasingly beleaguered Ottoman Empire, for example, Young Turks and Arab nationalists developed and advocated ethnic definitions of identity in an effort first to strengthen and eventually to undermine loyalty to the Empire on the basis of equality among Ottoman patriots. At the same time, a growing pan-Islamist movement argued that the Empire's troubles reflected a failure not to adopt European templates but to adhere to the tenets of Islam itself.

These same debates saw expression within the European Jewish community, and eventually gave rise to a nationalist movement,

---

[22] 'The Senatus-Consulte on the naturalization of Muslims and Jews in Algeria and the public administrative regulation for its execution, 14 July 1865–21 April 1866', cited in Hurewitz (ed.), *The Middle East and North Africa in World Politics*, 356.

Zionism, which encompassed both the ethnic nationalism and religious revivalism. Arthur Hertzberg nicely sums up the project:

> Until the beginning of the new age, the Jew conceived of himself as part of a holy community. . . . Since the Emancipation, Jewish thought has been attempting to rebuild a redefinition of Jewish identity, even with some – or many – of the bricks borrowed from the old building, but for a different need and from a different perspective: in order to make Jewish existence analogous to the categories by which western man has been defining himself. [23]

For Zionists, the tension between individual assimilation and collective identity was to be resolved by creation of a state that would be like all other states – the community rather than the individual would assimilate, and Jews would be able to live in that state as both Jews and citizens. To many early Zionists, most of whom were secular, indeed, irreligious Jews, that state could have been anywhere; for them, Jewish identity was an ethnic not a religious identity. The religious significance of Jerusalem and of the biblical land of Israel was an essential expression, however, of the religious impulse within the revivalists.

For Muslims, the tensions between individual rights and communal responsibilities were reflected in the varieties of nationalism that attended the demise of the Ottoman Empire. Some nationalists, including those who asserted control of what would become the republic of Turkey, borrowed heavily from Europe to construct their new state. With the abolition of the Caliphate in 1924, Turkey linked ethnic identity with the state and declared for a secular definition of its citizens, separating that state from religion. Elsewhere, nationalists struggled not only to construct comparable secular collective identities but to reconcile such ethnic identities – Arab, for example, or Kurdish – with the expectation that the states bequeathed them by the Europeans, which had been drawn for quite different purposes, would elicit patriotic attachments as well.

This left the Islamist reformists or revivalists at a loss, without any significant political expression for the next fifty years. Politics was

---

[23] Arthur Hertzberg, *The Zionist Idea*. New York: Atheneum, 1959, 21.

dominated by an ethnically defined resistance to European imperialism – Turkish and Arab principally but also Kurdish and, most recently, Berber in North Africa – that accepted the European-style state system even as it contested control, and even the boundaries, of the states themselves. As disenchantment with the governments born in the independence struggles set in, however, the radically secular nationalism they espoused began to tarnish as well.

Identity and community in the modern Middle East

For many Jews, Israel's victory in the 1967 war and its capture of the West Bank and Gaza seemed miraculous, fuelling religious enthusiasm and allowing a religious settlement movement to gather steam. Within the Arab world, that same event was profoundly disquieting; the search for solace contributed to the growth of Islamist movements that offered explanations and comfort. In both instances, the secular mission of the state eroded before a religious definition of political obligation and accountability. As they developed, Islamist movements came to represent not a demand for greater participation or better administration within the existing states so much as a genuine alternative to the state. By the 1970s, and certainly with the Iranian revolution, the once unimaginable restoration of the *umma*, the community of the faithful, became imaginable. By the end of the century, the notorious Saudi-born Islamist, Osama Bin Laden, reminded listeners in the broadcast in which he acknowledged al-Qa'ida's responsibility for the attacks of September 11, 2001, in New York and Washington that there was a long history to his grievances against the world:

> what the United States tasted today is a very small thing compared to what we have tasted for tens of years. Our nation has been tasting this humiliation and contempt for more than eighty years.[24]

Should his audience have missed the significance of the illusion to eighty years of humiliation, he clarified it several weeks later: 'Following World War I, which ended more than eighty-three years

---

[24] 'Broadcast by Osama Bin Laden (7 October 2001) in Barry Rubin and Judith Colp Rubin (eds.), *Anti-American Terrorism and the Middle East*. New York: Oxford University Press, 2002, 249.

ago, the whole Islamic world fell under the crusader banner – under the British, French and Italian governments'. In this broadcast, moreover, he identified the successor to the League of Nations as part of the problem:

> For several years our brothers have been killed, our women have been raped, and our children have been massacred in the safe havens of the United Nations and with its knowledge and cooperation. Those who refer our tragedies today to the United Nations so that they can be resolved are hypocrites who deceive God, his prophet, and the believers. Are not our tragedies caused by the United Nations? Who issued the partition resolution on Palestine in 1947 . . . ? Those who refer things to the international legitimacy have disavowed the legitimacy of the holy book and the tradition of the prophet Muhammad, God's peace and blessings be upon him.[25]

As the twenty-first century dawned, the very meaning and purpose of political authority, the very nature of the constituents and purposes of governments were contested across the region. The model of statehood and citizenship introduced by Europe was fast eroding, challenged by conceptions of obligation and accountability that transcended the frontiers of the territorial states.

For the Muslim world, the responsibilities to protect the holy cities of Mecca and Medina and to facilitate fulfilment of the religious obligation of pilgrimage for millions of Muslims every year, extended well beyond the formal obligations of the member state of the United Nations known as Saudi Arabia. The famed Egyptian Islamist, Sayyid Qutb, writing in the 1960s, seemed increasingly prophetic:

> If Islam is again to play the role of the leader of mankind, then it is necessary that the Muslim community be restored to its original form. It is necessary to revive that Muslim community which is buried under the debris of the man-made traditions of several generations, and which is crushed under the weight of those false laws, and customs which are not even remotely related to the Islamic teachings, and which, in spite of this, calls itself the 'world of Islam'.[26]

---

[25] Broadcast by Osama bin Laden (3 November 2001) in Rubin and Colp Rubin (eds.), *Anti-American Terrorism and the Middle East*, 256.

[26] Sayyid Qutb, 'Corruption's of the Modern World' (1964), cited in Marvin Gettleman and Stuart Schaar (eds.), *The Middle East and Islamic World Reader.* New York: Grove Press, 2003, 305.

This is a world that extends well beyond the confines of the Arab world, or even the region we call the Middle East, and in which authority and legitimacy are hotly contested. As Peter Mandaville recently observed,

> A country such as Malaysia, usually to be considered on the margins of Islam both in terms of geography and religious influence, has invested heavily in information and networking technologies. As a result, when searching on the Internet for descriptions of programmes which offer religious training, one is far more likely to encounter the comprehensive course outlines provided by the International Islamic University of Malaysia than to stumble across the venerable institutions of Cairo, Medina or Mashhad.[27]

Authority within the world Jewish community is similarly contested, as is apparent in debates about Israel's obligations as a Jewish state. David Ben Gurion, first Prime Minister of independent Israel was crystal clear in his vision of his constituency: 'The builders of the Land and the State are only those who dwell and live within it', he said, 'In the Diaspora, Jews as Jews are human dust, whose particles try to cling to each other'.[28] Ben Gurion's privileging of the inhabitants of the modern state of Israel was challenged in the aftermath of the 1967 war and the occupation of the territories associated with biblical Israel. The Jewish state was increasingly defined not simply as a state whose majority was Jewish but a state that transcended the immediate time and place and served all Jews. As Samuel Freedman points out, when Menachim Begin was elected Prime Minister in 1977, he 'emotionally embraced the concept of Eretz Israel, the Land of Israel, as defined by the Bible, not modern armistice lines' and the leader of the religious settler movement, Rabbi Tzvi Yehuda Kook, defiantly declared that 'Those who want to withdraw from Judea and Samaria will be cursed by the Almighty. . . . We are commanded by the Torah, not by the government'.[29]

---

[27] Peter Mandaville, 'Digital Islam: changing the boundaries of religious knowledge?' *SimNewsletter,* International Institute for the Study of Islam in the Modern World, March 1999, 1.

[28] Samuel G. Freedman, *Jew vs Jew: The Struggle for the Soul of American Jewry.* New York: Simon & Schuster, 2000, 167–70.

[29] *Ibid.,* 170.

By 2004, a group of religious members of the Israeli Knesset argued that Jews living around the world should be given the right to vote in Israeli elections. Rabbi Zalman Melamed argued in Jerusalem that 'there are 12 million Jews in the world. The state of Israel should be a state for the Jews of the world'.[30] Rabbi Melamed's proposal reflected a pragmatic concern that growing numbers of non-Jews in the territories controlled by Israel might jeopardize its claims to be both Jewish and democratic unless Jews elsewhere were accorded voting rights. Nonetheless, his proposal exposed the deep divisions within and well beyond Israel about the very principles and purposes of the Jewish state.

The proposal to extend political rights in Israel to Jews who are not Israeli citizens revived the long simmering debate over who, exactly, is a Jew. In the early days of independence, authority in religious matters, including conversion to Judaism, was assigned to the Orthodox leadership. Most American Jews are not Orthodox and they resent the privileges the Orthodox enjoy in Israel, and perforce, among Jews around the world. Like Islam, in which tensions over authority in the twenty-first century stretch from Jakarta to Dakar, challenging the venerable traditions of Medina and Cairo, within Judaism, similar tensions express themselves in a variety of Jewish communities. As Ismar Schorsch, the chancellor of the Conservative Jewish Theological Seminary in New York put it, in the debate over who is a Jew, 'Israel is the battlefield but the war is in America'.[31]

For Islamists like Bin Laden, America may be the battlefield but the war is in the Arabian peninsula, where Islamist complaints mix the pragmatic or worldly – widespread dissatisfaction with the performance of the Saudi regime – with aspirations for a restored *umma*, a political expression of Islam around the world. Bin Ladin and his fellow Islamists have argued for decades that the states of the Middle East – particularly but not solely the Saudi kingdom – are damaging to Muslims, dividing, embarrassing and weakening the community of the faithful.

---

[30] Mati Wagner, 'Give Diaspora Jews a vote', *Jerusalem Post*, 29 December 2004, http://www.jpost.com/servlet/Satellite?page-name=JPost/JPArticle/ShowFull&cid=1104291022152.

[31] Freedman, *Jew vs Jew*, 77.

FIGURE 5.3: Elective identities in a Jerusalem market
(© David Wells/www.davidhwells.com)

Perhaps not so surprisingly, there is a comparable debate about the value of Israel for the Jews. In an article that sparked considerable controversy in the autumn of 2003, a distinguished historian working in the United States suggested that

> The behavior of a self-described Jewish state affects the way everyone else looks at Jews. . . . The depressing truth is that Israel's current behavior is not just bad for America, though it surely is. It is not even just bad for Israel itself, as many Israelis silently acknowledge. The depressing truth is that Israel today is bad for the Jews . . . In a world where nations and people increasingly intermingle and intermarry at will; where cultural and national impediments to communication have all but collapsed; where more and more of us have multiple elective identities and would feel falsely constrained if we had to answer to just one of them; in such a world Israel is truly an anachronism.[32]

[32] Tony Judt, 'Israel: the alternative', *The New York Review of Books* **50:16**, 23 October 2003.

103

Lisa Anderson

'The continuing scene of political and economic turmoil'

Are the states formed by the borders drawn on maps of the Middle East anachronisms? Will the twenty-first century see polities divorced from territory, citizens with multiple elective affinities, novel definitions of public obligation and accountability?

The answers to these questions are being shaped today in the crucible of the conflicts in the Middle East. That they spill beyond the region's borders explains some of the ambiguity and uncertainty about the shape of the region itself. These conflicts about citizenship and faith, state and community, are battles fought in New York and Paris as much as in the Middle East. From suicide bombers in Riyadh to the attacks of September 11th, from the Jewish settlements in Gaza to the fevered campaigns against anti-Semitism in Europe and the United States, the battles over the definition of community are multiple, both in and beyond the geographical Middle East. Moreover, they divide communities, or what have been called 'civilizations', against themselves as much as they pit them against each other. Both Egyptian President Anwar Sadat, who was assassinated in 1982 by Islamists linked to groups which would later collaborate with Bin Ladin's al-Qa'ida in opposition to Egypt's peace agreement with Israel, and Israeli Prime Minister Yitzhak Rabin who was assassinated in 1995 by a Jewish fundamentalist opposed to the surrender of land he believed had been promised to Israel by God, died at the hands of co-religionists and fellow citizens.

Neither the West nor Islam, neither the states nor the cultures of the Middle East, are the settled, agreed-upon communities implied by Bernard Lewis or Samuel Huntington. Not only is the characterization of the Islamic world offered by Lewis and Huntington too simple and too monolithic to be plausible, they themselves disagree about the meaning of the civilization they so urgently represent and defend. The 'West' that confronts the bloody borders of Huntington's Islam is not quite the same as the 'Judeo-Christian heritage' that Lewis believes provokes Muslim rage. In fact, there was no 'West', at least in Huntington's terms, in the seventh century when Lewis' 'ancient rivalry' began, no real separation of spiritual and temporal authority, and no attachment to rule of law, social pluralism, representative

104

bodies or individualism. Moreover, for Huntington there is no *Judeo-Christian* heritage – his Western civilization is based on classical and Christian traditions alone. The competition to define the place of Jews in the West is reflected in the debates about Israel's place in the Middle East.

This Middle East – East of nowhere and in the Middle of everything – is a foil, a screen on which are projected the debates about the deep and painful fractures *within* all communities and civilizations at the dawn of the twenty-first century. That, in fact, is why they are so anguished. This is where imaginations run wild, where internal turmoil is played out, where the battles of our nightmares spill into the daylight, shaping the very real and too often deadly conflicts in the region. It is indeed, as the dictionary told us, defined by the fact that it is 'the continuing scene of political and economic turmoil'. How the conflicts in the Middle East are resolved will reflect and shape how lives are led – who we think we are and how we define our obligations – far beyond the fluid borders of the region.

FURTHER READING

Mitchell Cohen, *Zion & State*. New York: Columbia University Press, 1992.

Samuel G. Freedman, *Jew vs. Jew: The Struggle for the Soul of American Jewry*. New York: Simon and Schuster, 2000.

David Fromkin, *A Peace to end all Peace: The Fall of the Ottoman Empire and the Creation of the Modern Middle East*. New York: Henry Holt and Company, 1989.

Robert W. Hefner (ed.), *Remaking Muslim Politics: Pluralism, Contestation, Democratization*. Princeton: Princeton University Press, 2005.

Samuel P. Huntington, *The Clash of Civilisations and the Remaking of the World Order*. New York: Touchstone, 1996.

Mahmood Mamdani, *Good Muslim, Bad Muslim: America, The Cold War and the Roots of Terror*. New York: Pantheon, 2004.

6 Observing conflict

KATE ADIE

Former chief news correspondent, British Broadcasting Corporation

Years ago, as a member of the National Youth Theatre, I recall watching rehearsals of *Richard III* from the wings as the Alarums and Excursions of Bosworth Field were being conjured up by an over-enthusiastic scrum of lads laying about each other with sword and mace at the back of the stage. The director was bellowing above the din, 'Boys and girls, please do not kill each other. This is drama. No one need die.'

Conflict of all kind runs through Shakespeare's plays, many of the plots turning on it – emotional, historic, martial. He knew how to keep the audience's attention. And at the final curtain, the audience left the theatre and returned to a world which they knew was uncertain and shaped by war, but was incalculably different to the world of the stage.

Television also presents a kind of stage. We can witness both comedy and tragedy. However, it is as if the back wall of the stage dissolves every so often – and the real life that is walking past is thrust centre-stage, on-screen. Such is the technique of television that it is not always obvious what is reality and what is fiction. I have only to refer you to the constant enquiries to the BBC of those people who wish to have a drink in the Queen Vic pub in *EastEnders*.

For many decades, television has encouraged and honed various types of programmes to exploit and fulfil its possibilities: outside broadcasts and sport, intimate drama, series and soap operas, discussion and documentary, so-called 'reality shows' and so on. The medium is young and still developing; change is inevitable. News and current affairs were not originally thought to be pillars of the broadcasting schedule, but from the 1960s onwards enjoyed a phenomenal growth in popularity and esteem.

The journalist's fascination with conflict is well documented. In the nineteenth century in America, the Mexican and the American Civil War first began to awaken publishers to the fact that increased circulation accompanied war reporting. In Britain, in the 1850s, William Howard Russell of *The Times* almost single-handedly inaugurated modern British war reporting from the Crimea. Subsequently, there has been a bumpy road through the battlefields, encompassing censorship, bias, patriotism, propaganda and courage, eye-witness description, gritty realism and historic record.

Each major conflict brings its own particular characteristics to bear on the press and media: just as every general knows that the lessons learned *after* a war are always enshrined in doctrine to be dusted off and used for the next one – only to be found hopelessly out of date and misleading – every journalist should know that 'lessons learned reporting the last lot' will be peculiarly at variance with the realities of the 'next lot'. However, this did not stop an editor bombarding me with messages that it was essential that I secure a 'serviceable boat' when heading for a civil war a couple of years after the Falklands War. I gave up trying to explain that the Chad in Central Africa had no discernible coastline and I was not in search of the source of the Nile.

And in the second half of the twentieth century, with the new-born television medium completely ignored during World War II, we have only had just over fifty years to learn how we can report conflict with what is, for the moment, the most powerful communication tool in the world. My business has been mainly television reporting, so I would like to concentrate on that area.

And so quickly is television changing in itself and altering in its relation to the audience, it is a challenge to pin down how it reports conflict now – and how it should. I have lost count of the students who have contacted me in the last few years as they embark on their thesis about the military and the media. A goodly number are looking for a set of hard and fast rules about conflict reporting. They assume that there must be absolutes. All I can say is that, in my experience, the fog of war extends right through the camera lens and into the newsroom.

Let us try and peer through the mists of battle and discern some shapes which loom large in most conflicts.

*

I do not use the most common phrase much beloved of those who would disparage television. 'Dumbing down' suggests to me a conspiracy: a concerted attempt to deprive viewers of more intelligent and serious programming, under the guise of populism, accessibility and anti-elitism. I don't think there is any attempt to do this.

What seems to me to be happening is a swift move towards a more commercial view of the industry, after several decades in which – in this country – it had an automatic slice of serious and minority viewing embedded in the schedules, partly through regulation, partly through convention and tradition. Children's programmes, regional production, religious affairs, current affairs at prime time.

There is a 'lightening up' process in action. Culturally complex and serious current affairs and documentaries are being scheduled out of the main viewing hours. Programming style is leaning towards the more engaging; the overall move is towards the dominant strand of television entertainment. This has been happening for the past twenty years in American TV – and it's now the case in most of the developed world. Much of this is to do with ownership. The big deals in the media world are about profit and power; they are hardly ever about production values and creativity. Public service organizations from Canada to Australia, never mind the UK, are all feeling the pressure of the increased push for larger audiences, the competitive streak of commercial broadcasting. Television entertainment is now a highly exportable commodity; whereas for years only a few American programmes were promoted worldwide – I remember the oddity of watching *Dallas* in Bulgarian –programme 'formats', as they are now called, can be put together and sold very profitably as a template for production around the world.

But surely, news is exempt from these pressures? The news is the news. Why should it be exempt?

Nearly two decades ago I remember driving past large billboards in Los Angeles from which four impossibly glossy people with terrific teeth smiled engagingly and promised to bring me 'All the news you can use'. It'll never happen back home, I thought. In between the three major networks entertainment programmes, up popped little ads from the newsroom urging the viewer to 'join us for the latest'. Surely not, I thought. And newscasters were not distant, rather anonymous readers, but mega-personalities promoted vigorously by their stations.

It is the power of entertainment: the fact that television – unlike the theatre – deals with this extraordinary mixture of the real and the fictional, and entertainment values are inescapable in the hunt for audiences. Someone in the industry once said that television always wanted to smile at the viewer – after all, you've just been asked into their living room. However, the game show or soap opera will dissolve to give way to the news bulletin – or, to use the more recent term, the news programme, or now, the news show. And although the content is real conflict, real shock and surprise, genuine significance and sometimes unpalatable fact, the framework is that of the news presenter urging the viewer to 'Stay with us', a smiling 'Join us after the break' – as if the hard diet of news was something you'd like another helping of. And conflict is often on that menu. For most television news organizations still stick to an agenda of significant events as defined by broadsheet newspapers, and have not yet gone down the tabloid road – though there have been some experiments, as in BBC 3's *Liquid News*.

How to report conflict within this framework?

The tools available to a reporter these days are truly wondrous. The mobile phone and the satellite systems have worked miracles. Information can now speed round the world in a fraction of a second and sometimes deliver stories as they happen, in real time, as the expression is.

To give you an example: going to sea with the Navy up to a decade ago usually resulted in a reporter severing all links with the newsdesk – splendid stuff: you didn't get badgered by nervous editors, and they didn't have you whingeing every day. What communications the Navy had went to weird shore-stations and were labelled 'journalists, not for the use of'.

All that changed when the BBC trialled satellite equipment on board the aircraft carrier *Illustrious* in the Gulf during the no-fly operation in the late 1990s. With a little tweaking of the ship's course, we managed to broadcast live pictures from her flight deck.

However, during the 1999 Kosovo crisis, we were flown out to an American Cruise missile destroyer in the Adriatic; and because US ships have different electronic systems which tend to interfere with TV cameras and satellites, we expected only to be able to record limited pictures with difficulties – and would have to fly them off for transmission a few hours later. We stood on the deck near the ship's bow, the engines not turning, no light to be seen, as the ship waited for the moment of firing: the huge Cruise missiles were stored vertically in pods across the deck – not a great distance away from us. The signal to fire is generated not by the captain, but hundreds of miles away at a NATO headquarters in northern Italy.

Without warning, there was an explosion as the compressed-air apparatus pushed the missile out of the pod and up into the air in less than a second, before the rocket engine fired in a singeing blast and the missile hurtled up and curved towards the horizon. The stills photographer standing next to us pressed his button and hoped. He went below into the wardroom where he plugged his camera into his computer and was relieved to see that his picture was good: he cropped and captioned it – 'first Cruise missile fired at Serbia'. The computer was connected to his satellite phone, on which he dialled the computer layout of a New York newspaper. He pressed a key, and the picture was embedded in the front-page layout of the newspaper before the rocket hit its target.

It's obvious to anyone that live cameras and digital satellite communications systems challenge the elements of secrecy and surprise in warfare. The military are not unaware of these developments. They quite frequently see the press able to communicate faster than they can and they have concerns about operational security. Depending on the status of the press within a society and understanding of freedom of information, armies around the world vary in their reaction to journalists in a war zone. Add in the *casus belli* and the nationality of the journalist, and issues arise immediately of press freedom, patriotism, responsibility for fatalities and the right of people to know what is being fought for in their name.

The instinctive historic reaction is for generals to suggest that the press stay away and gratefully receive an official military report of victory at the end of the unpleasantness. At the other end of the spectrum, many of the public do not wish to know about the horrors of war.

Somewhere between these views are the majority of the nation and the press.

*

And I have mentioned 'nation'. For all the claims made for a 'global village', television news remains determinedly nationally based; the BBC and ITN and Sky News are British in outlook and ethos; CNN is a very American channel, though it does not wrap itself in its country's flag like FOX News. Al-Jazeera is based in Qatar and sees the world from a panArabic standpoint.

And so few journalists could or would claim a purely 'international' outlook, which raises the dilemma in times of conflict of, for example, loyalty and patriotism. Or, the espousing of one cause against another. Or support for the freedom-fighter. Or the reporting of terrorists' aims and objectives. Or encouragement for the human rights demonstrator. Reporting which calls into question the traditions of detached, objective, non-involved journalism.

It is so much easier to write despatches when you have decided that one opponent has right on their side. The cruelties and violence, the loss of life, the sacrifices, all can be more easily described when you do not have to balance your report. My country, right or wrong. For many journalists – though not all of course – consideration of objectivity goes out the window if your country declares war. Reporters – those seasoned, international, cosmopolitan creatures – discover that they have roots. And allegiances. And perhaps an editor or proprietor firmly behind the military venture. Or an audience fired up with patriotism. The idea that nationalism is something which belongs to history disappears overnight to the sound of drums and bugles, and gunfire.

Nothing is new in this. Reporters have always followed the flag, and it is a test of a country's liberal principles and freedom of expression if reporters' voices raised in dissent continue to be heard during the battle. And there may be a considerable problem for the press who wish to keep their audience informed of everything that is happening. A nation at war has no desire to see the press helping the enemy. And the satellite phone is certainly going to do that.

And the last fifteen years have seen considerable growth in the kind of conflict where one country intervenes in the affairs of another – not as an aggressor, but as an agent of the United Nations, or of regional groups. It became a regular observation in former Yugoslavia by journalists, that loyalty to the United Nations was an unknown emotion among the press. On the other hand, with dozens of journalists killed in that conflict, it had to be noted that the UN didn't kill them and so the question of loyalty was complex. And the actual complexity of the fighting, with at least three warring factions, produced very uneven journalism with the foreign press sometimes aligned with one faction, sometimes detached from all of them. Put crudely, some journalists had great difficulty reporting a war which didn't divide simply into good guys and bad guys. They could never decide what to do about the third lot.

Let me dispel one myth that features frequently in discussion about coverage of modern warfare: the idea that 'war is now seen "live" on television'. Only the occasional action is witnessed 'live' – and it is very limited and defined, for the moment. Even lightweight TV transmission equipment still needs carrying and running with and taking cover with in hostile situations. House-to-house fighting, civil warfare, battlefield close-quarter combat – these are situations in which the military neither tolerate media teams coming in among them nor can the teams operate coherently, though it may be possible in the future as developments deliver cameras and transmission equipment no larger than the weaponry carried by the infantry.

Set-piece attacks, on the other hand, favour the media: bombers taking off, missiles being fired, artillery in action, troop transports and armoured columns moving forward, can all be delivered to screens at home 'live' – as long as there is no sustained or overwhelming incoming fire from the enemy. This was the kind of footage which came in from the invasion of Iraq in 2003. And much of the 'noise of battle' – on analysis – turned out to be overwhelmingly *outgoing* fire, though it was mistaken for *exchanges* by many in the American forces at the time.

To be alongside in such situations raises the question of official tolerance of the media on the battlefield. In the Gulf, there had been a

decision at the highest political level to allow the media to be there –
under restrictions. Delivered in a manner which suggested an open-
ness and transparency – an invitation to witness the war – the so-called
'embedding' process was always intended to limit and control access
to much of the conflict. Additionally, there was definite disapproval of
any media who wished to operate independently; in the event, several
journalists were illegally arrested and detained by US forces for not
concurring with the official policy. The British media reported much of
the invasion from a neighbouring country – Kuwait, only being ferried
across the border when the military decided the time was right.

No one should be surprised at these strictures. Soldiers will always
feel an overriding need to protect their own troops from the consequences of uncontrolled publicity about their movements – and
detailed scrutiny of the viciousness and bloodiness of real fighting.
And politicians – ever more conscious of public opinion shaped by
media images – are nervous of taking responsibility for the ghastliness
of war and the sacrifice of civilians in its prosecution. You only have to
look at the language which has evolved to place a sanitized screen
between the brutality and the audience – 'collateral damage', 'smart
bomb', 'precision weapons' and so on – to realize that the realities of
war are unpalatable.

And if the set-pieces of attack are available 'live', there is also tolerance – up to a point – of set-piece defence. The Iraqi authorities placed
limitations on reporters, but allowed cameras on rooftops to deliver
pictures of distant flashes as bombs exploded. It suited the Iraqis to
have these images disseminated throughout the Arab world and to
those they wished to sympathize with them (Western audiences are
often unaware of the growing influence and sophistication of media
sources which are not based in their own capitals). However, an orange
glow on the horizon is an image which does not properly convey the
damage done by modern weaponry. And dying and death were notably
not on screen.

*

The arguments as to what is acceptable on that small-screen stage in
the living room reflect the culture within a community. There is no universal norm for the amount of violence and extent of suffering which

can be tolerated on television. Conflict presents particular problems in this respect: should the full impact of violence appear on air? Shredded corpses and screaming victims? Fighters delivering death-blows? Napalm and fragmentation weapons hitting civilians?

The borderline between what is acceptable and what is not is regularly shifting. Even during a conflict there are inconsistencies: dead Allied soldiers were treated with the respect of distance when shown. The mangled corpses of Saddam Hussein's sons were considered OK in close-up.

Even in the Bosnian conflict in the early 1990s, where there was very limited live coverage, the increasing speed and internationalization of TV coverage began to impinge on the highly sensitive area of military information about casualties being communicated before their own authorities had informed relatives at home.

It's usually considered proper by most countries that the families should be informed properly and privately of a death. In Sarajevo a young French soldier was shot in the head while building an anti-sniper barricade. His death was recorded by an agency cameraman who'd been watching the barriers being put up. There were close-up pictures. Less than ten minutes later, the time it took to drive down Sniper Alley in the centre of that city to the local TV station where the satellite dishes were located, the pictures were transmitted over the Eurovision network. That was how the soldier's family learned of his death, before he had even been pronounced officially dead. All conflicts now – even civil disturbances and riots – have the potential to involve families and friends while the action is in progress. Those families and friends may well subsequently constitute a considerable influence as to the conduct of the conflict. However, other than an on-the-spot appeal to the camera crew, there is little the military can do, and it's reasonable to expect these days that much of frontline action is recorded by agency crews rather than national organizations.

It also has to be noted that armies in the West are now mostly volunteer, and relatively small compared to the forces of World War II. This leads to the military being more isolated from the majority of society, and fewer viewers feeling a direct link with those putting their lives on the line in a foreign land. A poll in America last year showed that two-thirds of those surveyed had no relative, friend or colleague serving in Iraq. The structure of the US army and its recruitment also lead to a disproportionate number of black and Hispanic Americans serving, along with poor white Southerners. So the link to the forces which might be assumed in the majority of TV viewers is weaker than usually supposed. Add to that the Administration's dislike and discouragement of coverage of returning body-bags and pictures of wounded troops in hospital and rehabilitation – all essential elements of reporting conflict and its consequences – and there is a reinforcement of the notion that the 'entertainment-led' medium does not find it too difficult to avoid certain aspects of journalism.

*

There is the underlying argument that really appalling images will chase the viewer away – for good, so destroying the *raison d'être* of the news. Coupled with this is the fear that continuously grisly pictures eventually produce indifference. And the suggestion that if the news shows nothing but carnage, then it produces a generalized misanthropy, as Michael Ignatieff has said, 'the feeling that the world has become too crazy a place to deserve serious reflection'.

All of this has been exacerbated by the arrival of twenty-four-hour news. It is replacing that sense of occasion which used to occur when the newspaper dropped on your mat at a set time in the morning, and in the evening you made an appointment with the serious half hour of the nightly news bulletin. Instead, there is now a continuous stream of variable information – some of it factual reporting, now interspersed with comment, opinion, discussion and a presentation style that owes more to advertising than to lecturing.

Conflict – of all kinds – does not run according to television scheduling patterns. I tuned in at a very early hour during the Iraq invasion and watched a camera near an infantry dug-in position trained on half-a-dozen men lying in sand, doing very little. The horizon showed no action. The reporter had a hard time finding much to say. One of the presenters in the studio memorably turned away from the back-projection screen with the words, 'Seems there's not much going on in the war this morning – let's leave it and look at today's sports prospects'.

Such coverage can trivialize and also minimalize conflict, suggesting that cameras are surely placed at all the main events and big theatres of action – which is, of course, not true. But the emphasis given to 'going live' tends to reinforce the idea.

*

Twenty-four-hour news also puts immense pressure on reporters to deliver something new and different far more frequently, usually from a so-called 'live-spot'. That, in essence, has to be a relatively safe location, in which it then becomes possible to deliver a large number of reports to various outlets, updated every quarter of an hour, if desired. A highly economical use of resources, but one which restricts the eye-witness reporting capacities of the reporter. It has led in both Afghanistan and Iraq to reporters rarely, if ever, leaving the building and relying on news copy sent by e-mail from Washington or London. The reporter then delivers information gathered by someone else. The frequency of the broadcasts can also lead to the over-emphasizing of minor events, so that it becomes difficult for the viewer to evaluate the course of the action. And the course of the action, the conduct of the conflict, is now delivered with corporate-style gloss by Western authorities: press conferences, selective video clips and photo opportunities are as much part of the official military information process as any other business.

All this sounds rather pessimistic for the pursuit of facts and significant information. On the contrary, the facts are probably there in greater profusion than ever before. It is just that they come in a welter of other material, and the viewer is hard-pressed to evaluate them and weigh their significance. Surely there is no obligation to watch twenty-four hours of conflict? However, does it seem right just to 'drop in' on a war, when it suits you? The relationship of the viewer to the television news is a little uncertain in this respect. Nor is the viewer given an opportunity to understand the process of what is being censored on grounds of taste: warnings such as 'Viewers may find some images disturbing' do not inform the viewer of the height of violence that may have been reached, or the depth of suffering – and which may have been excluded.

Another plus should be that we are getting information a great deal faster, however confusing it may be. There is an argument that a democratic society should be kept informed, not just told the result at the end of the affair. Decisions are being taken by those involved in the conflict, and it may be that the public develops a view that such decisions should be questioned, perhaps changed. The Vietnam War is often cited in this respect, though it is something of a red herring, the traditional argument being that pictures of dead and injured GIs had such an impact on nightly news broadcasts that a groundswell of opinion eventually dissuaded the administration from pursuing the war. This was not the case – there was no noticeable growth in opposition to the war while the largest number of casualties were being taken. The pictures had no measurable effect on public opinion. The anti-war movement had different origins.

Again, on the positive side, the media in the West have never been better funded and so prolific: an explosion of TV channels, with work in the media seen as attractive and fashionable. Few people who have never encountered the media circus when it hits town seem to have much idea about the sheer number involved: gathered in Skopje in Macedonia before NATO troops went into Kosovo, a British army officer was curious to see how many journalists he was trying to deal with. He set about registering them – and journalists are suckers for a badge with the word 'Accredited' on it – not that it ever confers rights and privileges, it merely identifies you as 'Not a soldier'. He ran out of time and knew he hadn't got everyone, but he already had 2,734 on the list. The state of Qatar reckoned that more than 8,000 media arrived to attend the US military's briefing centre during the Iraq invasion. Somewhere in that lot there has to be a smattering of excellence, brilliant journalism, fair-dealing, honesty and accuracy?

And perhaps that's what's on the mind of journalists heading for conflict – those tricky moral issues which lie at the heart of the trade: the pursuit of truth, fairness, accuracy and lack of bias. Are we all there amid shot and shell, pondering these matters? Probably not, we are trying to stay alive. One of the consequences of improved technology is the longer time which the press are present in conflict. There is no need to leave the battle-front to find the telephone or send the pictures. The satellite transmission equipment goes as near as it can to the scene of the action – though not too near to sustain a hole in its expensive dish. So, it is not surprising that the number of press, dead and injured in the last decade, worldwide, has increased.

The figures for this last year from the Paris-based *Reporters without Borders* show fifty-three journalists and fifteen media workers killed, at least 907 journalists arrested, 1,146 attacked or threatened and 622 media censored. Of course, many of these cases were reported in non-conflict areas; however, Iraq led the list, with nineteen reporters and twelve media workers killed during the year. Terrorist strikes and Iraqi guerrilla attacks caused most of the deaths, but the US army was held responsible for four of them. What this points up is the increase in deaths since President George W. Bush stood triumphantly on the deck of an aircraft carrier to signal a war all but won – and the enormous media bandwagon trundled home from the Middle East. With a conflict becoming much more complicated to report, less of a simple traditional two-sided fight, it has also become more dangerous and less attractive to the world's media.

And there's no doubt that the availability of cheap, lethal automatic weapons, particularly since the break-up of the Soviet Union, had added to the dangers. There is also an awareness, almost as world-wide, that television in particular can be used with great effect in a conflict – not just to report it, but as propaganda, a weapon of war itself. So the business of staying alive is a serious one. I had never seen a flak-jacket worn by a reporter – never mind worn one myself – until the Balkan war. And up to the 1990s, I doubt that any reporter had driven an armoured vehicle. Sadly, young reporters now take these things for granted.

And there is an added element today which places restrictions on even the most independent-minded young reporter: employers are now highly sensitive to the pressures from the Health and Safety industry and also insurance demands. What has been standard practice for years among journalist in conflict areas is now being modified: head offices and newsrooms are increasingly risk-averse, nervous of compensation litigation and worried that death or injury will reflect badly on corporate image. One effect is positive: better-equipped journalists – though it can also be argued that armoured vehicles and flak-jackets can distance the media from other civilians. However, the net result of administrative caution is the tendency to accept offers of 'protection' – from the military or private organizations – with an eagerness which diminishes the opportunity for independence.

There is also an insidious pressure which arises from the ever-rising graph of fatalities: the call for journalists to be covered by some sort of International Code of Ethics, which would be intended to give them some kind of protection in conflicts. The suggestion usually comes from governments who are none too happy with foreign journalists crawling over a particularly nasty little war of civil insurrection. If only they could identify the pesky critters – then they could afford them the correct 'help and attention' and perhaps some 'protection'.

In my experience, away from the relatively sophisticated older members of NATO, the first question asked by a general, indeed, by most people in uniform, is: Who regulates you journalists? Behind this lies the universal military desire to establish a chain of command. It is anathema to soldiers that journalists are such free spirits; that their behaviour cannot be supervised by a code of practice. I fear that an International Code of Ethics – which isn't much discussed in this country but is a favourite topic in countries with a little local difficulty in progress – would be seized upon by most men with guns: not to give journalists freedom to operate, but to restrict and to control.

The military mind will seize on any code which contains a hint of the words 'internal affairs of a nation', 'in the national interest', or 'territorial integrity' – or even fairness and balance. You can imagine how the military interpret such words as aiding and abetting the enemy, especially when lives are at stake. And as commercial and publicly funded organizations in the West become more cautious in their approach to dangerous assignments, let us hope that they do not even toy with the notion of a Code of Ethics: it would be corrupted before it ever got signed up to.

*

One comment on another development which also touches on the emotional and moral values we attach to involvement in conflict, and which television can emphasize disproportionately: a fighter pilot is shot down in battle; a prisoner of war is taken: routine story? Not if that fighter is a woman.

119

It's a straightforward fact that women live in the shadows of conflict. The history of war and insurrection, crime and domestic trouble, scenes of violence, though littered with the odd heroine, the occasional pioneer, the eccentric female, and the exception to her times, the oddball woman, is the story of man. Or men. Decisions on declaring war, the leadership in battle, the composition of armies, the maintenance of law and order in civil society have merely reflected the second-class status of women. Their lack of political or economic clout. And at times, the wilful exclusion by historians of redoubtable deeds by women, just because women weren't supposed to do them. No surprises there, and that's history for you.

However, in the twentieth century, there was an explosion of information, a massive growth in education, and whole populations now learn about the world in a way which was impossible, inaccessible, in previous centuries. Add to that the way in which the Western world has seen – in just two generations – the age-old assumptions about women's status challenged and changed.

Women do have views on conflict – and they may be varied – not just the conventional view that they are always pro-peace and anti-war. However, these views are frequently discounted, or not even heard, because the very presence of conflict makes it harder for women to break into the macho circle of decision-making.

I saw numerous example of this in the Balkans, where well-educated women, the beneficiaries of an efficient non-discriminatory Soviet education system, were elbowed into the sidelines as soon as the war began. Many were municipal officials, civil servants, media personnel. Their roles were gradually usurped in a militarized society, where wearing a uniform and boasting about front lines became the badge of success. Time and again, I attended gatherings in which ceasefires were negotiated, where refugee problems were discussed. They were exclusively male. I met many able and experienced former civil servants – these women expressed frustration and bewilderment at the way in which they'd been pushed aside with the crudest of arguments. War's not for women. Generals don't want skirts around. Your job's at home – the men are on the front lines – someone's got to look after the kids; anyway, the water's been cut off and there's no electricity.

One of the first signs of women's trek backwards in Sarajevo was the sight of middle-class women lawyers and teachers and architects lugging pails of water and chopping down trees for firewood. Their jobs seemed secondary in a war – so they had to take on the physical tasks thrown up by the damage caused by war.

A good number of them tackled their new situation with a resentful determination. One of the commonest sights in the middle of Sarajevo at the height of the shelling was that of schoolchildren clutching their brightly coloured bags, and darting across streets vulnerable to deadly sniper fire. One of the main secondary schools was on the riverbank in the north. It was a few hundred yards from the Serb front line. The headmistress wasn't a particularly tough character, but she explained, while wrapped in a duvet against the freezing weather blowing in through the four-foot mortar hole in her office, that war didn't stop education. And she was insistent – and I met many women to attest to this – that school was continuing because the pupils' mothers were frantic that their children would have a future – something other than fighting. It was the mothers, she said, who scraped together a bit of sawdusty bread and some ancient jam for the tiny lunch-box, and shoved their children out onto the streets to make their way through mortar and artillery fire and past snipers to school. Their argument was that as women, no one took any notice of what they said about the war, so they were going to make sure that their children at least got an education so that they could leave and prosper elsewhere. And anyway, they added, you were as likely to be shelled in your own living-room or kitchen as in a classroom, so off you go to school.

So again the women had lost their public voice because of conflict.

Another aspect of reporting conflict is to observe the way women have become involved in the actual business of fighting. The change in women's status in Western democracies in the last few decades has led to this.

Of course, there are numerous historical examples of females in combat. But it's interesting to note that the arguments about women in the front line are often accompanied by some of the hoariest old myths about women and conflict. One of the commonest is that men fighting next to women are likely – nay definitely – going to lay down their arms, abandon military discipline and forget what they're fighting for, the moment the woman next to them is injured. Chivalry.

I find this fascinating. Because chivalry is not something that seems to have gained ground in other areas where women have demanded equality. It's much more likely to have originated in the arguments that took place in the Israeli army over thirty years ago, where women had gained significant participation in the armed services, only to encounter objections from the religious element, who then had them ejected from front-line duties, a ban which has remained in place until the past couple of years.

However, many military planners in the Western world are now grappling with the role of women in the armed forces, not so much because they are yielding to notions of equality, but because they face a considerable shortfall in recruitment, added to the increasingly technological aspects of war, where better-educated people are needed. The traditional cannon-fodder – the uneducated, unskilled male – is no longer the major constituent of a modern army, navy or airforce. And most countries are having to widen their recruitment base because many of the qualified people they need happen to be women.

It's a complex situation, if only because of the fundamental emotions it arouses in everyone about the roles of men and women in war and peace, on the front line and the home-front. As a reporter, I have faced the inevitable pressure – which comes of convention – that the 'exceptional' constitutes a story. In other words, when you're working on the deck of an American aircraft carrier during full battle operations – a frightening, hellish place, with dangerous jet aircraft manoeuvring feet from you, while you climb over piles of rockets and bombs, all the while unable to hear anything because of the scream of the take-offs – there is the moment when the F-18 pilot waves from the cockpit before hurtling off at several hundred miles per hour from the catapult on deck. And she is a stunner. With a long blonde plait stuffed into her hi-tech helmet.

Worth a story? Or just another F-18 pilot?

And if something happens to her, then the media suddenly hurtle back several decades, deeply traditional and discriminatory in their view of a woman's place, leaving front-line reporters mulling over whether they have dealt with the 'exceptional' or merely reinforced old prejudices.

*

So, more media, better equipped, able to broadcast faster; wrestling with traditional conflict problems of military restriction. censorship and unpalatable pictures. Communicating to a television industry running twenty-four-hour channels, hungry for material, presenting itself ever more in entertainment mode.

What does the viewer make of this? If I had the answer, I'd be setting up my own station.

What is immediately obvious is that the pattern which lasted for nearly forty years, of ever-increasing audiences spread across a handful of channels, with a very discernible set of viewing habits, has now crumbled. And despite the proliferation of satellite and cable options, the audiences for any one particular programme are, with occasional exceptions, much smaller than a decade ago.

Programming which deals with the subject of conflict – documentaries and current affairs – have declined on terrestrial stations and are rarely to be found in peak-time viewing hours. However, this is countered with specialist channels – Discovery, History and so on. News has expanded on all fronts; even so, the time when nearly half the adult population could be expected to watch one of the main evening bulletins has gone. The twenty-four-hour news channels claim 'cumulative' audiences – adding together figures for various times, regardless of whether these may be the same people tuning in; and they are still only measured in fractions of the terrestrial channels. The fall-off has run parallel with a decline in newspaper readership. There are a variety of reasons put forward – society is changing, young people do not belong to the loyal generation which 'grew up' with television; news is considered less significant within busy lives and where there is no overarching international threat, such as the Cold War, also the 'emotional, touchy-feely' and personalized style of reporting grates with older viewers – and many other suggestions.

Nevertheless there's no doubting the ability of certain stories and images to reach out and affect public opinion. If images of conflict were not influential, then the military and political machines would not bother to attempt to control them.

There is cause for optimism. We have large numbers of young people interested in and caring about reporting; not all merely want to be 'rich and famous' – which is a much-voiced mantra. The equipment grows ever more sophisticated and can take the reporter into the conflict to deliver what he or she can see and show. The audience is not only important in terms of audience size. In a sophisticated and grown-up society there are large numbers of people who understand that conflict must be reported and reality confronted.

Long may we try to serve that understanding.

FURTHER READING

K. Adie, *The Kindness of Strangers*. London: Headline, 2002.

K. Adie, *Corsets to Camouflage: Women and War*. London: Hodder & Stoughton, 2003.

K. Adie and T. Grant (eds.), *From Our Own Correspondent*. London: Profile Books, 2005.

K. Adie, *Nobody's Child: Who Are You When You Don't Know Your Past*. London: Hodder & Stoughton, 2005.

7 Conflict and labour

WILLIAM BROWN

Department of Economics, Cambridge University

Conflict and labour are inextricably linked. But the link is not straight-forward, and nor are the symptoms of conflict. This essay is concerned with how both the link and the symptoms have changed. And the change has been very substantial. Strike levels are currently remark-ably low in historical terms, both in Britain and worldwide. But despite this, I shall argue that the significance of conflict to the relationship between employer and employee is undiminished. Indeed, the impli-cation of my argument is that the fall in strikes is itself a matter for concern.

Let me start by emphasizing two sharply contrasting aspects of labour. The image of industrial conflict with which anyone over the age of about thirty will be familiar is one of rowdy picket lines, locked factory gates, indignant banners and even police in riot gear. But it is now twenty years since the bitter national coal-miners' strike ended. That year-long dispute, which split and broke a once-powerful trade union, both symbolized and hastened the end of a period of British history during which overt labour conflict was rarely out of the head-lines.

Contrast that with a far more familiar employment scene in Britain in the twenty-first century, to be found, for example, anywhere in the Fens of Eastern England; one of recent migrant workers, tending, picking and packing vegetables. They come from all over the world: Lithuanians, Poles, Chinese, Brazilians, Portuguese, Kurds and so on. Some are directly employed by the farmers, more are employed through gang-master agencies. Some are illegal immigrants. They are typical of what is changing most in the world of labour worldwide in the twenty-first century. Uprooted, un-unionized, often unsure of the

125

language, and certainly of their rights, they are acutely vulnerable to exploitation by unscrupulous employers.

Many of these men and women are probably fulfilling the aspirations that brought them to the region. They are earning well above the Minimum Wage in full-time jobs, settling down in Kings Lynn and other local towns, and integrating into our increasingly diverse society. But they are the fortunate ones. The grim aspect of this new style of employment was tragically highlighted by the twenty-three Chinese immigrants who died cockle-picking in Morecambe Bay in the north-west of England in 2004. They would have been earning derisory amounts of £1 or £2 an hour after deductions, subject to blackmail and physical intimidation both here in Britain and for their families back home. They would be living a squalid and miserable existence, far from what they had been led to expect. Theirs is an extreme case of a widespread contemporary form of covert labour conflict.

These are the two faces of labour conflict I want to explore. On the one hand, the *overt conflict* of organized labour defending jobs and terms of employment. On the other hand, the *covert conflict* of unorganized labour unable to achieve what society would accept as decent standards of employment. But to set the scene, it is necessary to explain why conflict and labour are unavoidably linked.

Labour and conflict

For most people, for most of history, earning a living has been hard, unremitting, life-dominating, and usually unpleasant. For the great majority of workers, who have no special skills or labour market scarcity, labour is hard and always has been. Bear in mind that in developed societies such as ours, over the past hundred years, technological change has transformed work through steady increases in labour productivity. It has permitted the reduction of the length of the average working week by roughly twenty hours – a third. It has halved the proportion of that time necessary to earn the money sufficient to keep a worker alive – from over four-fifths to around two-fifths. And that is for a much-improved life – almost double life-expectancy at birth, three inches more height, and working not only shorter hours but with more energy because of better nutrition. Until only the past very few

FIGURE 7.1: Covert conflict: Above: Chinese immigrants cockle-picking in Morecambe Bay in the north-west of England in 2004 (© Dominick Tyler); Below: migrant workers vegetable packing in eastern England in 2004 (© author).

generations in developed societies, work dominated life far more than it does now; it was more physically exhausting, and it was driven by naked hunger.

For our society this is largely history. But it is still the lot of the developing world. Full-time workers in Britain today earn on average about £80 a day; half the workers of the world as a whole – between 1 and 2 billion people – currently earn less than £1 a day. And a great many earn a great deal less than that. Most of the workers of today's world experience terms of employment no better than those of our great grandparents.

The second point I want to stress is how *open* the employment contract is, both in terms of what the employee is expected to do – how long and how hard to work – and in terms of what the employer pays for that work. The central issue for this essay is essentially this: how are the labourer's load, and the labourer's reward, determined? If it is a bargain, it is certainly not in any sense a balanced one. The labourer may have skills and energy, but the employer has the land, the equipment, the contacts and contracts, and everything else that allows those skills and energies to provide a living.

Employees have always been weak. Rural labour in Britain was typically hired for a year, and if a worker tried to leave during that year, for whatever reason, the employer could call on the magistrates and constables to prevent it. Indeed, in nineteenth-century England there is a demonstrable inverse relationship: when unemployment fell, prosecutions for breach of contract rose and vice versa. For most of history, for most rural labour, the threat of prison alternated with the threat of hunger. More generally, for all of history, and for most of employment now, the relationship between employer and employee has been – and is – fundamentally unequal.

The third point I want to stress is that there has always been a tendency for some employers to abuse this extreme imbalance of power. When labour is abundant and has no rights and no means of redress, these abuses can be extreme. We have accounts by survivors such as Primo Levi and Alexander Solzhenitsyn of how the managements of forced labour camps were literally willing to work their labour to death. And this is not just a matter of history. It is cautiously estimated by the

International Labour Organization (ILO) that there are some 27 million people working in some form of slavery today, defined as being forced to work and being unable to leave because of physical intimidation. It is likely that there are currently more people in slavery than at any previous time in human history. Compared with the past, fewer of these are chattel labour; more are bonded and debt-enforced labour, and many in state labour camps. What is unchanged is that the conditions of such enforced employment are generally unimaginably horrible. The key point for my argument is simply how gross can be the abuse of the worker if the employer is unconstrained.

If tens of millions of workers are in slavery, many more – hundreds of millions – are kept at work for low pay and in harsh conditions not by force, or by contrived indebtedness, but by sheer necessity to earn a living. These include around 110 million children under the age of fifteen whom the ILO considers to be working in unacceptably dangerous or unhealthy conditions.

This is contemporary covert conflict. There may be no organized resistance, or even public protest, but the conflict of interest between the hard-pressed worker, and the employer who profits from their work, is so very extreme. The covert conflict arises to the extent that terms of employment fail to meet the best interests of the worker. There can be no objective definition of this but, at minimum, there would be general agreement that there is covert conflict when the employer gains at the expense of the health and well-being of the employee.

I am not, of course, arguing that employers necessarily take advantage of their overwhelming position of power. Far from it. There have always been progressive employers who, for both moral and economic reasons, treat their labour with respect. In many societies this is the norm. There is a long and honourable history of employers who can afford relatively good pay because of the productivity gains that come from good training, low turnover and high morale. Good employers are to be found in the developing as well as the developed countries.

But good employers find their task more difficult if the competition they face comes from unscrupulous employers, from employers who are uninhibited in taking advantage of the imbalance of power.

Winston Churchill summed up the problem succinctly when he introduced Wages Boards nearly a hundred years ago. First he set out the
moral position: 'It is a serious national evil that any class of His
Majesty's subjects should receive less than a living wage for their
utmost exertions. . .'. Then he noted a fallacious economic belief: 'It
was formerly supposed that the working of the laws of supply and
demand would naturally regulate or eliminate that evil. . .'. And he concluded with the sad reality: 'But where you have no organization, no
parity of bargaining, the good employer is undercut by the bad, and
the bad employer is undercut by the worst'.

The constraint of employer power

What constraints can be placed upon 'the worst' employers? Over the
past century or so, two quite different sources of restraint have developed. The first is *trade unions*, voluntary associations of workers using
their co-ordinated resolve to improve and protect the terms on which
their labour is sold. The second is *governments*, establishing and
enforcing both union rights to organize and also minimum standards
of employment – on minimum wages, health and safety protection,
maternity entitlements, penalties for unacceptable discrimination and
so on. Such rights are generally called 'minimum labour standards'. I
shall consider what has happened to these two basic ways of managing labour conflict – trade union activity and government provided protections – in turn.

You may find it odd to describe trade unions as means of managing
labour conflict. Are not they behind so much overt conflict? Well in
part, yes. The early days of trade unionism were in many ways
extremely conflictual. Certainly they were in Britain. The 1890s saw on
average something like a hundred times more working days lost per
year through strikes per trade union member than the 1990s. Canada,
Australia, Sweden, Denmark, Belgium, Russia were other countries
which experienced economic paralysis by strikes as the nineteenth
century turned into the twentieth century. Trade unions had to fight
for recognition by both employers and governments.

But recognition they got, with varying degrees of enthusiasm on the
part of employers and of legal support on the part of governments. The

establishment of the International Labour Organization in 1919, to which almost all countries became affiliated, blessed the view that membership of an independent trade union was a right in a democratic society. Very quickly an increasingly comprehensive coverage of agreements developed. They were agreements between associations of employers for a particular industry reached with the trade unions of their employees. These set out the minimum wage rates, maximum hours of work and a range of other details of employment for that industry.

This pattern of industrial agreements developed in most European countries in the early twentieth century, despite very different legal systems. It made sense because, within any particular country, the employers of an industry were mainly competing with each other. If they could reach agreement on basic pay and hours for their whole industry, they could offer decent living standards to their employees without fear of being undercut by a rival who could get away with paying less. It was called 'taking wages out of competition'. Even firms outside these industries, and untouched by trade unions, were brought within this patchwork of minimum labour rights. So long as the main competition for the goods and services came from within a given country, this provided a very stable system of industrial governance. It established national labour standards. And it did so in a way that was responsive to the varying fortunes of different industries.

With this came the widespread acceptance of trade unions. Their coverage and influence grew. In Britain, it grew sharply during the two world wars, partly as a result of the constructive role unions played in the war effort. At its peak around 1980 over a half of all employees in Britain were members of trade unions. And because the patchwork of collective agreements laid down basic wages and hours irrespective of whether one was a union member, by that time over three-quarters of all employees were protected by some sort of collective agreement. But after 1980 union membership collapsed to below 30 per cent of employees, and it has not recovered. The fact that membership is currently stable at this lower level is misleading. If we look behind the figures we see a deeper weakness.

In order to explain this I need to point out that trade unionism has always had two distinct aspects. On the one hand, the aspect of

FIGURE 7.2: Work stoppages and work days lost per union member in
Great Britain (1893–2003)

confrontation, whereby the union mobilizes collective worker strength
to bargain – to try to win or defend standards of employment. On the
other, the aspect of *co-operation*, whereby the union works as a channel
for communication, representing employee interests and helping to
facilitate change.

Confrontational trade unionism

In their *confrontation* mode, trade unions win and defend their gains
by strikes or threats of strikes – through overt conflict. Figure 7.2 gives
the history of strike propensity of trade union members since the
1890s. After the early tempestuous years, strike propensity fluctuated.
But it then fell in the past twenty or so years to the present historically
remarkably low levels.

The levels of disputes of the post Second World War years were very
high, and strikes were constantly in the news. But it is important to
stress that they were very localized. Strikes in the huge mining industry, with its coal-face-based piecework payment systems, dominated
the mid-twentieth-century statistics. And later it was again disorderly
pay systems and poor management that made, notably, the car-

assembly industry, docks, and ship-building industries conspicuous. The great bulk of trade union members never experienced a strike, although their employment was managed with the awareness of their strike potential. Organized labour was usually influential without being disruptive.

A recurring theme of the lectures in this series has been gender differences in conflict, which warrants two observations here. The first is that women's propensity to join trade unions overtook that of men in Britain in 2004; the proportion of women employees in unions has been fairly stable while that of men has been falling. The second is that, if you allow for workplace size and industrial structure, there is no evidence that women differ significantly in strike-proneness from men. Organization and overt labour conflict are not male preserves.

Co-operative trade unionism

There has always been a *co-operative* aspect of trade unionism. In this, the union acts as a channel for communication and consultation between the employer and employees, giving early warnings of potential problems, and giving approval to innovations felt to be in employees' interests. Employers typically value the legitimacy of endorsement that comes from the independence of trade unions. This is one reason why trade unions have generally been seen to make a positive contribution to industrial governance. Another positive managerial aspect of trade union activity is simply that it keeps employers on their toes. The evidence is that the management of employees is generally better where unions are present. It is a reflection of a wider observation that, throughout history, the availability of cheap, docile labour has discouraged technological and managerial progress.

Trade unionism also has to be seen in a wider perspective, as representing the interests of labour as a whole, whether unionized or not. One aspect of this has been the achievement of trade unions throughout the democratic world in successfully campaigning for governments to provide a floor of employment rights. Most recently in Britain this has been through the 'social partnership' role of the Trade Union Congress (TUC), given the task jointly with the Confederation of British Industry (CBI) of developing most employment policy since

1997. Another aspect has been the extent to which trade unions provide a parallel channel of citizenship, of democratic involvement through very immediate representative structures, including the shop steward organizations still central to much representation in British employment.

The gravity of this wider political role was reflected in the importance attached by the Allied Powers in 1945 to providing trade unions with a secure legal basis in the reconstruction of Germany and Japan, as a deterrent to renewed militarism. In the 1980s, it was reflected in the central role in the campaign for democracy played by *Solidarnosc* in Poland and Congress of South African Trade Unions (COSATU) in South Africa. In valuing this political role we should never forget the dangers of trade unionism where employers and government are hostile. The International Confederation of Free Trade Unions, which monitors these things, estimates that in 2002, 129 trade union activists were murdered worldwide; 213 in 2001. There are powerful and brutal people opposed to labour organization in many parts of the world.

The collapse of collectivism

Why did strikes and trade union membership decline so sharply in Britain towards the end of the last century? Much of the immediate explanation lies in better management. Employers have become more proactive and less reactive. They have become far better at preventing problems, with well-thought-out systems of payment, training, grievance handling, consultation, promotion and so on. Part of this increased professionalism reflects the advisory work of the Advisory Conciliation and Arbitration Service (ACAS), which also deals with around a thousand industrial disputes a year, defusing and resolving the vast majority by conciliation and, for a small minority, arbitration. Overt conflict is greatly reduced by good management and by the orderly use of disputes procedures, including conciliation.

What about all those anti-union and anti-strike laws of the 1980s and 1990s under the governments of Margaret Thatcher and John Major? These certainly did indeed increase the costs of striking and of trade union organization. Probably more important than the legislation was the willingness of those governments to back the employers during the set-piece strikes of the 1980s which challenged and broke the unions involved – in steel, newspapers, television, telecoms, shipping, rail and, of course, mining. But while this government action accelerated the decline in trade union influence in Britain, it is arguable that the legislation was pushing on an open door.

If we look at aggregate industrial unrest worldwide, we see that Britain's experience of a sharp fall in the past twenty years largely reflects that of other industrialized countries. Overt industrial unrest was peculiarly distinctive to the twentieth century. In a way the epic battles of the American steel unions at Homestead in 1892 and of the British miners' union at Orgreave in 1984 symbolically bracket the beginning and end of this conflictual period in the history of industrialization. The fight with striking steel workers taught American employers that it might be better to talk with unions than try to eliminate them. By contrast, the miners' strike rubbed home to British trade unions the ultimate futility of resisting a collapsing product market – in this case, cheaper sources of fuel than deep-mined coal. The collapse of both strikes and trade union membership since the 1980s has been common to most developed countries. Something mightier than Mrs Thatcher lies behind this international change. What is it?

It has been the increased internationalization and expansion of the world's economy that has been the driving force behind all this; behind more effective labour management, and behind weakened trade unions, worldwide. I mentioned earlier how the basis of labour standards for most of the twentieth century was the patchwork of industrial agreements which effectively 'took wages out of competition' in national systems of bargaining. But, the more the national market was undercut from abroad, the harder it became to sustain this nation-based system of industrial governance.

The increased internationalization of the world economy – globalization – applies both to *where* things are bought and sold and to *who owns* the firms that buy and sell them. If we take the period 1980 to 2000, while world output has increased on average at 3 per cent per year, world trade has increased at double that amount, 6 per cent. An

FIGURE 7.3: Overt conflict. Above: the American steel unions at Homestead in 1892; Below: Orgreave Coking Plant, near Sheffield, 19 June 1984 (© Martin Shakeshaft).

ever-growing share of what we buy comes from another country. As a result, employers have been increasingly unable to stick to old nation-ally-based agreements because their newer competitors are overseas. The problem for trade unions is that, despite endless attempts, they are unable to organize collective action across national boundaries. Worker solidarity effectively stops at the frontier.

So far as *ownership* goes, the average rate of growth of foreign direct investment per year was 12 per cent – twice that of world trade and four times that of world output over the period. Fewer and fewer private sector workers are ultimately employed by firms based in their country. Owners increasingly wander the world, choosing to invest where the labour-cost suits them best. And this is typically in countries where trade unions are weakest. For unions, the viability of strike action is sharply diminished when they know that a successful strike may simply hand the jobs to workers in another country.

To add to trade union woes, their position has been undermined by the increased exposure of public sectors to competition. Unions that were once strong in sectors such as government, education, health, and public utilities have seen their membership and influence eroded by privatization and commercialization. Despite public services being sheltered from world markets, growing pressure from tax-payers has had similar consequences.

More recently, to this increasingly globalized world economy, there has been a further, historically unprecedented shock. Quite simply, the global economy has doubled. The post-war capitalist world that traded substantially with each other – the Americas, Western Europe, the Asian Tigers and so on – amounted to about 1.5 billion economically active people by the end of the twentieth century. But after about 1990, very suddenly, we saw the collapse of communism in Eastern Europe, the rise of market socialism in China, and the emergence of India from protectionism as a serious competitor. The number of economically active people engaged in competition with each other has sharply doubled to 3 billion. Much of the medium- term competitive threat to the established trading world from these 'new globalizers' comes from their cheap labour. But the longer-term challenge comes from the high priority the new economic arrivals give to education. A third of college students in the world at the moment are in these newly arrived countries. Increasingly their challenge will come from higher productivity, not lower wages.

There is no reason to expect a reversal in these trends. Employees are increasingly exposed to competitive forces against which traditional trade union methods are impotent. That is why strikes have fallen worldwide. That is why they are unlikely to return to past levels. And it means that our once-settled (if disputatious) system of fixing labour standards, and of defending workers from unscrupulous employers, has been eclipsed. In Britain, for example the coverage of collective agreements collapsed from covering three-quarters of the workforce in 1980 to about one-third today.

Does this mean that trade unions have lost their role? It certainly means that the confrontational role that so characterized the mid-twentieth century has been greatly diminished. Instead unions are increasingly seeking co-operative relationships with employers. In Britain, this can be seen in the growth of workplace partnership agreements in which unions typically offer employers, on the one hand, greater flexibility of working and the foregoing of strikes in return, on the other hand, for better training, consultation and job security.

It is not only that the coverage and influence of trade unions is diminishing. The same economic pressures are making it difficult for trade unions to cover the weakest and the least skilled in the labour market, whose jobs are particularly exposed to international competition. Throughout the world, trade unions are increasingly becoming confined to the professions, the skilled, and the directly employed public sector, now more of a 'labour aristocracy' than when the term was coined in the nineteenth century. Unsurprisingly, this weakening of trade unions is reflected in the wider trade union movement. The TUC's influence with government, so great from the 1940s to the 1970s, is greatly diminished, except in so far as it can work with the employers' counterpart, the CBI.

Does this mean that working people are left defenceless in the twenty-first century? Are jobs once again exposed to the cruelties of market conditions as they were until the late nineteenth century? Are we fated to see covert conflict over labour spiral upwards?

138

Government intervention

Let me turn to the second restraint on employer power that I mentioned: government intervention. Governments have traditionally
intervened in two ways. The main way for most of the last century was
by giving trade unions legal support, to help indirectly to uphold the
patchwork of industrial agreements I have described. But more
recently the emphasis has been on governments giving individual
workers protective rights, rather than relying on trade unions to nego-
tiate agreements to do this.

In Britain in the past thirty or so years employees have gained many
individual protections: against unfair dismissal; for health and safety
at work; for redundancy compensation; against sex, race, disability and
age discrimination; minimum wages, maximum hours, holidays, parental leave and so on. These reduce the scope for exploitative treatment.
The state is providing a substitute for some of what trade unions once
fought. These new protections are undoubtedly both needed and effective. I shall illustrate this with reference to the British minimum wage.

The National Minimum Wage was introduced in 1999 at the end of
a period during which, as Figure 7.4 shows, the relative position of the
low paid had been worsening, in part because of the economic trends
already outlined. Perhaps the worst implication of this deterioration
was for child poverty: between 1980 and 1995 the proportion of children living in households with incomes below half the average income
rose from 10 per cent to 30 per cent. Since its introduction, the
Minimum Wage has benefited around 1.5 million workers. It has
scrunched up the bottom of the income distribution. The people at the
bottom have received, as Figure 7.4 shows, substantial and continuing
relative gains. Child poverty has diminished substantially. Here we see
that government intervention has, in a very short time, provided effective protections that trade unions on their own could not.

There are several ways in which the management of the Minimum
Wage is relevant to the protection of labour in this new era of weak-
ened trade unions that I have been describing. First, the Low Pay
Commission that runs it is independent of government, and (like
ACAS) is a social partnership body. Its nine members include three
trade union representatives even though very few of the employees

William Brown

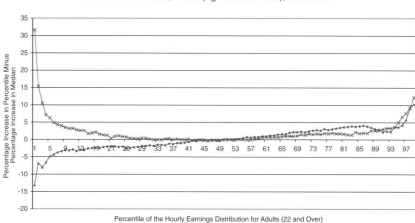

FIGURE 7.4: Increase in percentile gross hourly earnings excluding overtime minus increase in the median, adults (aged twenty-two and over), 1992–2003

directly affected by the Minimum Wage are themselves trade union members. Second, the Commission consults very widely, going out and about to explore the grubby end of the labour market. It is also informed by unusually authoritative commissioned research. And third, unlike other individual rights, the Minimum Wage is enforced by the Inland Revenue, whose rat-catching zeal in hunting down non-compliance is awesome. Around £3 million in back pay is restored to cheated workers each year. At least in this case, government intervention to uphold labour standards appears to be successful, with social partnership steering energetic investigation and enforcement.

How far can we expect governments to protect the weak? Can they deliver? The problem is that governments are no more in control of global markets than are trade unions. Let me illustrate this with the example of the length of the normal working week. It has always been a highly symbolic issue for trade unions. It was noted earlier that, over the course of the twentieth century, the length of the working week had fallen, broadly, from sixty to forty hours. Each step-by-step reduction (without loss of weekly pay) was seen as a triumph for workers every-

140

where. The last national step – M. Jospin's introduction of the thirty-five-hour week in France in 2000 – was one the most ambitious.

But international competitive pressures are challenging this century of solid progress. At the government level, the French thirty-five-hour week is currently being undermined by legislation to increase the scope for exceptions. Perhaps the clearest threat lies in the extent to which it is not so much unorganized workers, but the most organized workers with elite employers of Europe who are challenged. In 2004, the union at Siemens' North Rhine-Westphalia telephone plant was confronted with the threat that 2,000 jobs would be moved to Hungary. Following demonstrations the union agreed that the working week should be increased from thirty to forty hours, with no increase in pay. At much the same time, the French Bosch factory near Lyon was faced with the threat that all jobs would be moved to the Czech Republic, where labour-costs were less than half as much. The workers voted to work an extra hour for no more pay. There could be no clearer example of, first, the hard-won achievements of trade unionism being eroded; and, second, of the inability of national government to provide alternative protection.

International labour standards

Globalization has pulled the regulation of decent labour standards beyond the reach of both trade unions and nation states. What, if anything, stands between us and ever-deepening inequality worldwide? On the one hand, profits are being channelled to insulated enclaves of privilege around the world; while on the other, we have what is often called the 'race to the bottom' – with governments underbidding each other in the labour standards they enforce in an effort to win investment and retain jobs. Can the creation of some sort of international labour standards stop this downward race?

The creation of the International Labour Organization (ILO) at the end of the First World War arose from just such concerns. Shocked by the Bolshevik revolution, the governments of the then-developed world were fearful that, if their economies were undercut by cheap foreign labour, unemployment might provoke similar revolutionary feeling at home. Since then the ILO has been zealous in drawing up international standards – on trade union freedom, against forced labour, slavery, child labour and so on – and in getting the bulk of the nations of the world to subscribe to them. The sad fact is, however, that without powers of enforcement, little has come of this. Slavery, child labour and the persecution of trade union activists, as we have seen, continue unabated. Efforts to use organizations such as the World Bank, or the World Trade Organization to enforce international labour standards are opposed by the developing world. They see such restrictions, with good reason, as thinly veiled protectionism by the developed world.

The developing world also sees such restrictions as not being helpful to those they purport to help. Take, for example, efforts to enforce international bans on child labour. Most children work because their families need the money – denying them jobs simply increases their families' hardship. Far more helpful for those children are positive measures to help their families. A growing number of developing countries have recently brought in policies whereby families are paid to ensure that their children go to school. This is not a restriction, but an investment to improve their children's futures. This is the sort of action that is making a difference.

Recently there have been significant achievements through consumer campaigns. Within Britain, consumer pressure is having a substantial impact. The big supermarket chains have become acutely sensitive to the damage to their reputations that comes from revelations that their suppliers are bad employers. They have, in conjunction with the TUC, established a system of Ethical Trading Initiative auditors who check on the employment conditions, legal status of workers, National Insurance payments, wage levels and so on of supplying firms. In a very short time these have become a substantial regulatory influence.

The same is happening internationally. Revelations in the 1990s about appalling labour standards of manufacturers supplying, for example, Nike, Gap, IKEA and Wal-Mart have forced these firms, that rely so much on brand image, to institute their own factory-inspection routines for their suppliers. The results are readily available on their websites. Similarly, the Fairtrade movement offers guarantees that the money paid for tea, coffee, bananas or whatever actually gets to the farmers who do the physical work. Consumer campaigns, feeling their way up the supply chain, have become an important weapon for trade unions as well as for broader ethical interest groups. In a world where multi-nationals are often greater than nations, and electronic communication is ever easier, consumer campaigns seem to be achieving more than ILO conventions.

But even effective consumer campaigns can only have a limited impact. The deeper solution to the challenge of labour exploitation lies in political action. Covert labour conflict can only be reduced if the gap between rich and poor is narrowed. Until recently, the most blatant gap was between developed and developing countries, between North and South. But much of the growing inequality of the world now lies *within* countries. It is true that average incomes in some countries, especially in Africa, continue to fall away ever further from the developed world. But for the new big players in the world economy the gap between their average earnings and those of the old developed world is narrowing. Where inequality is deepening is *within* these countries. The key to diminishing covert conflict in these new arrivals lies within themselves.

I started by saying that conflict and labour are inextricably linked. I argued that there is always a potential for conflict, whether in the overt form of strikes, or the covert form of gross exploitation. Just as overt conflict can be reduced by good management and by good bargaining procedures, so covert conflict can be reduced by reducing inequality. And what will reduce inequalities within countries will not depend on the imposition of labour standards from outside. It will depend on the health of internal democratic processes, processes that force politicians and employers to take not only their common people seriously, but also the trade unions that represent them.

FURTHER READING

W. A. Brown, 'Industrial relations and the economy', in R. Floud and P. Johnson (eds.), *The Cambridge Economic History of Britain, Volume III, 1939–2000*. Cambridge: Cambridge University Press, 2004.

R. Donkin, *Blood, Sweat and Tears: the Evolution of Work*. New York: Texere, 2001.

R. W. Fogel, *The Escape from Hunger and Premature Death, 1700–2100*. Cambridge: Cambridge University Press, 2004.

8 Life in a violent universe

P. C. W. DAVIES

Australian Centre for Astrobiology, Macquarie University, Sydney

During 2005, scientists working with the Chandra X-Ray Observatory reported detecting the biggest cosmic explosion ever. A galaxy situated 2.6 billion light years away has been ravaged by a monstrous black hole that has swallowed the equivalent of 300 million suns in an orgy of destruction that has been going on for 100 million years. This ingested material isn't going quietly. As the scattered debris plunges to oblivion within the black hole, it releases vast amounts of energy, creating a pair of high-speed jets that have blasted twin bubbles 650,000 light years across in the host galaxy. You wouldn't want to be living anywhere nearby that monster!

Violence is the leitmotif of the universe. It was born in a big bang. Its fundamental structure was forged in the first split second, in a searing maelstrom of unimaginable ferocity, at temperatures exceeding a trillion degrees. Its history is one of cataclysmic explosions, implosions and collisions of literally astronomical proportions, of titanic forces and enormous energies. Yet amid this cosmic mayhem, life has not only emerged, but flourished – at least it has on one planet. How has something so delicate and precious as biology made a home amid the chaos of a violent universe?

The take-home message of my essay is that violence has a creative as well as a destructive aspect, and that without exceedingly energetic and powerful processes that seem so awesome to human beings, life would be impossible. To illustrate this message, I shall draw upon three examples of astronomical violence in ascending order of severity.

Asteroid and comet impacts

The biggest external threat to life on Earth today comes from the impact of comets and asteroids. This is a familiar hazard since the release of the Hollywood blockbuster movies *Deep Impact* and *Armageddon*, and scarcely a month goes by without the press running a scare story about the latest impending cosmic collision which turns out to be a false alarm. Over the past decade or so, work by a handful of astronomers has led to the detection of hundreds of potentially Earth-intersecting asteroids, and it is now possible to produce a reasonably accurate measure of the threat we face from a chance collision. The figures can seem alarming. Expressed most starkly, the chances that a randomly chosen individual will die from an asteroid or comet impact this year is greater than one in a million. This is more than the chance of being struck by lightning, and comparable to the threat of death from an airline disaster.

Asteroids mostly orbit safely between Mars and Jupiter, but some of them get disturbed and sent towards the inner solar system where they pose a threat to Earth. Their destructive power stems not so much from their size as from their speed. Travelling at twenty to thirty kilometres per second, an object just one kilometre across hitting a planet can release more destructive power than all the world's nuclear weapons put together. Objects this size slam into our planet on average every million years or so. If one hit tomorrow, it would most likely fall in the ocean, raising tsunamis a kilometre high around the ocean rim. If it struck the land it would excavate a crater twenty kilometres across. The blast from the explosion would wreak havoc in the immediate vicinity, but the long-term damage of dust in the atmosphere and wild fires caused by molten rocks plunging to the ground across the globe would prove the greatest danger, possibly by creating nuclear winter conditions. The disruption to crops and economic activity would kill many more people than the actual impact. It is hard to tell how many would die from an impact of this magnitude, but estimates of a billion or more are not uncommon. So we are presented with a very small probability of a very big catastrophe.

As there are many more small asteroids and comets than big ones, the chances are higher that we will be hit by something the size of

Darwin College. This happens on average about once a century, the last recorded occasion being the famous Tunguska explosion of 1908, which flattened trees for thousands of square kilometres in a remote area of Siberia. The Tunguska event was the equivalent of a ten-megaton nuclear bomb. Mostly these smaller explosions occur above ground, but from time to time the incoming objects hit the ground and excavate a large hole, such as the famous Meteor Crater in Arizona, or the Wolfe Creek crater in Western Australia. Needless to say, if this sort of event occurred over a major city, the death toll could run to millions. Of course, most of these objects fall over the ocean or wilderness, but it is only a matter of time before a large population centre takes a hit.

The impact of a one-kilometre-wide asteroid anywhere on Earth would be an appalling disaster, but it would be unlikely to wipe out humanity. However, still-larger impacts happen from time to time. Every 10 million years or so, on average, a ten-kilometre-sized object will strike the Earth. Such an event has the potential to annihilate entire species. Indeed, the fossil record shows several sudden episodes of mass extinctions, in which a large fraction of extant species abruptly disappears. It is not clear to what extent one or a spate of cosmic impacts might have been responsible in each of these episodes. The most famous case is the impact 65 million years ago that is held by some to be responsible for the demise of the dinosaurs. It left a 180-kilometre-wide crater in what is now the Yukatan region of Mexico.

The plethora of asteroids milling around the solar system constitutes the remnants of detritus left over from the formation of the planets. The ferocity of the bombardment – which afflicted all the planets and moons of the solar system, not just Earth – was far worse in the past. For about 700 million years after the planets formed, bodies up to several hundred kilometres across menaced the inner solar system. We can tell from looking at the moon, which preserves the record of a horrendous barrage, that dozens of huge impacts occurred up until about 3.8 billion years ago. Earth, being a bigger target, would have taken more punishment. The biggest impacts would have blasted a crater the size of the British Isles, stripped away much of the atmosphere, and swathed the planet in incandescent rock vapour and super-heated steam at a temperature of 3000°C. Under these hellish conditions, the oceans boiled and the Earth's surface was sterilized. The traces of ancient terrestrial craters are mostly obliterated by weathering and tectonic activity, and only a few dozen are known. For this reason, the role of impacts in shaping Earth's history was overlooked by scientists for many centuries, although as long ago as the seventeenth century Sir Edmond Halley speculated that comets would strike the Earth from time to time.

Fortunately there is negligible risk of a mega-impact today, but there remains a persistent threat from asteroids and comets with size-range up to ten kilometres, which can still cause enormous damage. Slowly governments are waking up to the threat posed by cosmic collisions, and there is now some effort being expended to track down the most dangerous asteroids to determine their orbits. That way we can work out when – and it is when, not if – the next one will hit. Very little can be done, however, about long-period comets, which literally come out of the blue in an unpredictable manner. Nobody knows what, if anything, could be done if we had notice of an impending collision. In principle, an approaching object could be deflected to a near-miss orbit. For this to be successful, we would need many decades of warning. Blasting the object with nuclear weapons, Hollywood style, wouldn't be the answer, for it would merely multiply the problem by turning a bullet into buckshot. Rather, controlled nuclear explosions near the object's surface might nudge it onto a new orbit. This strategy would depend on whether the asteroid is a solid object or merely a heap of rubble held together by its own weak gravity. Comets seem to be rubble and dust glued together by ice. But we don't really know. In July 2005, a NASA mission called Deep Impact involved striking comet Tempel 1 with a copper projectile travelling at a relative speed of over 10 kilometres per second. Analysis of the plume material released from the impact site should yield valuable clues about the internal constitution of the object.

It seems to me that it will be a long time before humanity enjoys what Arthur C. Clarke has termed 'spaceguard' protection – an early warning system of approaching objects which can then be systematically intercepted and dealt with, somehow. Meanwhile, it would be prudent for us to establish a permanent human settlement on Mars. Over a few centuries, this outpost could be made self-sufficient, and would serve as a lifeboat for our species and culture in the event of a global catastrophe here on Earth. Until we have either set up a permanent self-sustaining colony beyond Earth, or developed the technology to anticipate and attend to all future incoming asteroids and comets, there will remain a small but significant probability of human annihilation from a cosmic impact.

So far, I have dwelt on the destructive aspect of impacts, but true to my theme I will now present the good news. I have already mentioned the death of the dinosaurs. Biologists think this event cleared the way for mammals to predominate, leading eventually to the emergence of *Homo sapiens*. Without the Yukatan impact, we wouldn't be here today. That's good news for us, bad news for the dinosaurs. Evolution is like that – winners and losers – as Darwin taught us. Destruction of one species can speed the emergence of others. And there is no bigger eco-disaster than a cosmic impact.

My colleagues at the Australian Centre for Astrobiology have been studying another major impact that occurred 578 million years ago, in what is now South Australia. Lake Acraman on the Eyre Peninsula is now known to be the remnant of a huge impact, the age of which correlates rather well with the so-called Cambrian explosion. This term refers to the sudden proliferation of complex biological species after 2 billion years of little more than microbial life on Earth. My colleagues conjecture that the severe environmental stress occasioned by the Acraman impact propelled the diversification of species, under the imperative 'Adapt or die!'

It now seems fairly clear that large impacts have played a major role in shaping the history of life on Earth by eliminating species and acting as a motor for evolution. There is a hint of fine tuning here. Too many impacts would imperil all complex life-forms, but too few would result in a static biosphere with little incentive for innovation. It is unlikely that sentient beings like ourselves would emerge on a planet that suffered many more or far fewer impacts than does the Earth. It is interesting that the rate of impacts is determined in large measure by the planet Jupiter, which acts as a sort of gigantic vacuum cleaner, sweeping up incoming comets and flinging many of them out of the solar system before they can hit the inner planets. No Jupiter, and the impact rate would be lethal.

Impacts played an even more creative role in the origin of life. Soon after the Earth formed it was struck by a Mars-sized object. The outer layers of both planets were stripped off and coalesced to form the moon. (Some people think the moon has played an important role in the story of life too.) This awesome encounter desiccated our young planet, leaving its surface barren and hostile to life. In the ensuing several hundred million years, the arrival of comets laden with ice and organic material cloaked the sterile Earth with a veneer of life-encouraging substances. Amazingly, there were enough comet impacts to fill Earth's oceans many times over. Liquid water is indispensable to life, at least as we know it, and we owe it to comet impacts that our planet is two-thirds covered with water.

Although these early impacts established fertile ground in which life could take root, it is far from clear that terrestrial life originated here. Sterilizing impacts from the heavy bombardment period continued until about 3.8 billion years ago. My colleagues have obtained, and continue to obtain, credible evidence for life in the rocks of Western Australia dated at 3.5 billion years ago. Since life is highly unlikely to spring into existence ready-made, its origins must stretch back well before that time, which makes it seem as if life established itself on our planet almost as soon as conditions permitted. This suggests that it came to Earth from somewhere else, and for my money that somewhere else is Mars. Being a smaller planet, Mars cooled quicker. It avoided the cataclysm of a moon-forming impact. Its lower gravity meant that it rapidly lost the hydrogen released by the photo-dissociation of water, leaving the oxygen to rust the planet – hence its red hue. These events created favourable physical and chemical conditions, such as an ozone layer and redox gradients (roughly speaking, a supply of chemical energy), to drive the sort of metabolic cycles that primitive life-forms need.

Although Mars offered a better environment for life to get going, Earth had the rosier long-term prospects. Low gravity condemned

Mars to turn into a freeze-dried desert by 3 billion years ago, as its atmosphere leaked away and the temperature plummeted. Fortunately life had an escape route! Violent impacts ejected copious quantities of Martian material into solar orbit. Any microbes living inside the ejected rocks would have been conveyed into space. Although space conditions are very hostile to life, cocooned inside a rock, microbes would have been protected from the lethal radiation. The cold, dry conditions of space would have preserved them very effectively, especially if they had adopted spore-like forms. Thus protected, life could have remained viable in space for millions of years.

The significance of the foregoing escape route is twofold. Solar orbit could have offered a temporary refuge for Martian microbes after a big impact. Some of the ejected material would eventually have fallen back, recolonizing the red planet with its own life-forms once conditions had returned to normal. The low gravity of Mars made this orbital refuge readily accessible. Secondly, some of the ejected material eventually fell to Earth. We know this for a fact, because a couple of dozen Mars rocks have been found as meteorites. Extrapolating from these finds, it is clear that billions of tonnes of potentially fecund Martian material has plunged to Earth during its history, much of it in the crucial period 4 to 3.5 billion years ago. If life did start on Mars, say 4.4 billion years ago, it would almost inevitably have spread to Earth by this mechanism, and established itself quickly once the heavy bombardment abated. So cosmic impacts not only 'fertilized the ground' for life on Earth, they may even have delivered the product to a waiting planet. Without impacts, life would have remained stuck on Mars, its prospects bleak.

Exploding stars

For my second example of a serious hazard to life on Earth I shall discuss the explosion of stars. Clearly such an event represents a much higher level of violence than cosmic impacts, but the effect on our planet is offset by the vast distances to the stars, so that as far as mass extinctions are concerned the threat works out about the same. Indeed, there seems to be a scaling law at work in the universe that reduces the probability of a violent event (equivalently it increases the expected distance away from us of the next such event) in proportion to the energy released, implying that the level of threat to Earth is comparable, whether it is an asteroid impact right here or the collision of two super-massive black holes several billion light years away. If the universe did not have this scaling law – if more violent events happened more often (increasing the probability of them occurring closer to Earth) – then intelligent life may never have emerged.

Stars explode for a variety of reasons, but I shall consider here only one, the so-called type-II supernova event. This involves stars somewhat more massive than our sun, but there is no lack of such objects in our galaxy and others. Stars contain huge nuclear reactors in their innards, and so long as the heat keeps coming the stars keep shining. The internal temperature generated by nuclear fusion creates enough pressure to support the star's immense weight, but eventually the star runs out of fuel and faces a crisis: it can no longer generate the heat needed to shore itself up. In these circumstances the core of the star may collapse catastrophically under the force of gravity, releasing a huge amount of energy. As a result, the outer layers of the star are blasted into space at high speed. For a few days this titanic explosion causes the star to shine billions of times more brightly – hence the term 'supernova'.

Much of the energy is released as a pulse of neutrinos – ghostly subatomic particles that normally enjoy stupendous penetrating power. For example, neutrinos from the sun continually hit the Earth – and pass right through our planet with hardly a shudder. But the imploding core of a star is so dense that even neutrinos find it hard to make progress. As a result, their outward rush is resisted, and the ensuing force drives much of the infalling material back out again, to create the supernova explosion. A typical type-II supernova releases the energy equivalent of at least 10^{30} hydrogen bombs, which would spell calamity for any orbiting planet. Certainly no life could survive on such a body.

Fortunately our sun is too small a star to blow up in this manner, but even a supernova in our general region of the galaxy poses a threat. And it does happen. There is accumulating evidence that a supernova occurred in our galactic vicinity about 3 million years ago. Rocks from the seabed contain isotopes of iron with a short half-life, indicating they could not have formed with the solar system. Iron is thought to be a major component of the gases dispersing from a supernova explosion, suggesting that Earth was engulfed in the cloud of debris from a dying star.

The main threat to life from a nearby supernova explosion comes not from any spray of material, but from the accompanying intense burst of gamma radiation. Fortunately Earth's atmosphere affords a measure of protection in this respect, and it has been estimated that only if the exploding star lay within ten or twenty light years would serious biological consequences ensue. There is currently no star in that range likely to become a supernova, but the threat doesn't stop there. Stars of very large mass can produce prodigious gamma-ray bursts when they die by concentrating the released energy into narrow beams. Although the details are not fully understood, and involve strong electromagnetic as well as gravitational processes, what seems to happen is something like this. The core of the star collapses to form a rapidly spinning black hole. Material swirling down the hole releases huge amounts of energy that becomes collimated into narrow jets on opposite sides of the black hole. Some of this energy ends up as gamma radiation, and any planet lying in the path of the jet would be catastrophically irradiated, even if it were hundreds or even thousands of light years away. There is a dangerous-looking very massive star called eta Carinae about 10,000 light years away that is currently spewing forth copious quantities of gas and could blow up at any time. In 1843 it was seen to flare up and was briefly the second brightest star in the sky. Fortunately there is no reason to expect Earth to lie in the firing line of the ensuing beam of gamma rays, should this explosion happen soon.

Supernovae are undoubtedly dangerous and destructive, but they also play a key positive role in the story of life. Some astronomers believe that a supernova triggered the birth of the solar system in the first place, a suspicion gleaned from the presence of certain isotopes in meteorites, thought to have been present when the solar system formed 4.55 billion years ago. Chinese scientists have recently identified within a so-called carbonaceous chondrite the isotope sulfur-36, which they interpret as the decay product of chlorine-36, a radioactive isotope with a half-life of just 300,000 years. The most plausible source of the chlorine-36 was detritus from a nearby supernova that occurred within 300,000 years of the formation of the meteorite, thus implicating the supernova in the birth of the solar system.

A more significant role for supernovae in the story of life concerns carbon, the life-giving element. Carbon was not made in the big bang; rather, it has been manufactured over billions of years in the cores of stars. If stars didn't explode, or divest themselves of material in other ways, this carbon would be confined to stellar furnaces and not made available for organic chemistry to work its life-producing magic. Every supernova explosion seeds the interstellar medium with all manner of chemical elements, not the least of which is the carbon, nitrogen and oxygen so vital for life. As Sir James Jeans once remarked, our bodies are made from the ashes of long-dead stars.

My colleague Charles Lineweaver has studied the evolution of habitability within our galaxy throughout its 13-billion-year history. Initially the galaxy was an unpromising system for life. The central regions contained a high density of large mass stars that burned rapidly and produced frequent supernovae in a relatively crowded region of space. On the other hand, outside this region there were too few elements such as carbon for life to flourish. After about 2 billion years, however, a goldilocks zone began to open up in the galactic suburbs, a region in which there was enough carbon to make life and few enough local supernovae to destroy it. Over time, this 'galactic habitable zone' has expanded somewhat. Our solar system is located near the outer edge of the current zone, and at 4.5 billion years of age, it is about one-third as old as the galaxy.

The big bang and the multiverse

As a final example, I now turn to the explosive origin of the universe itself: the so-called big bang. Roughly speaking, the big bang was equivalent to 10^{55} hydrogen bombs detonated simultaneously! The searing heat from this origination event still bathes the universe in the form of a cosmic microwave background – a fading afterglow at a modest temperature of just under 3000°C. But in the past this radiation was hotter. At one second, it was about 10 billion degrees; before that, hotter still. Obviously no life could have existed in the immediate aftermath of the cosmic birth. Quite apart from the extreme temperatures, there was no carbon. Astronomical evidence suggested it took several hundred million years before the universe cooled enough to permit the first large stars to form, which then exploded to seed the interstellar medium with life-encouraging substances.

Whilst a searing explosion may seem an unpromising start for a universe destined to spawn life, a closer examination reveals it to have been peculiarly bio-friendly in several respects. The big bang was not just a chaotic, energetic outburst, but an exquisitely orchestrated affair. If the explosive vigour had been slightly greater, the cosmic gases would have dispersed too rapidly for galaxies, stars, planets and – presumably – life to form. But had the bang been weaker, it would not have overcome the mutual gravitation of all the cosmic matter, and the universe would have collapsed back on itself in a big crunch before stars had time to form. Either way, the universe would probably have gone unobserved.

For a long while cosmologists were baffled by the manner in which the vigour of the big bang was so felicitously matched to the gravitating power of the universe, but today they explain this happy coincidence by appealing to the so-called inflationary universe scenario. In this theory, the universe is deemed to have abruptly leapt in size by a stupendous factor in the first split second, propelled by an intense pulse of antigravity that created a fleeting phase of high-speed runaway expansion. The dynamics of this overwhelming antigravity are such that, within the twinkling of an eye, the universe was automatically propelled on the path of expansion at precisely the rate needed to avoid a big crunch, yet slow enough to permit the local aggregation of matter to form galaxies. In other words, a life-encouraging universe is built into the basic cosmic dynamics, if the inflation theory is right.

Related to this issue is the fact that, on a very large scale of size, matter is distributed remarkably smoothly across the universe. Also, the expansion rate of the universe is extremely uniform. These facts are important for life, because a universe that expanded rapidly in some regions or in some directions, and slowly in others, would be dangerously chaotic, leading to the disruption of galaxies by collisions. Inflation neatly explains this uniformity, because the antigravity that kick-started the cosmological expansion was totally democratic: it singled out no favoured direction or cosmic region for special treatment. Thus any initial irregularities would have been smoothed away. Rather like inflating a balloon, with enough stretching it becomes highly uniform.

On the other hand, had the universe started out with exact uniformity, matter would never have aggregated into galaxies and stars. Fortunately, the large-scale smoothness of the early universe was accompanied by (relatively speaking) small-scale irregularities – variations in the density of matter of about one part in 100,000. These tiny perturbations show up in a thermal map of the sky recently compiled by a satellite called the Wilkinson Microwave Anisotropy Probe, or WMAP. The radiation imaged by WMAP has travelled almost undisturbed since about 380,000 years after the big bang, and is effectively a snapshot of what the universe was like at that early stage. Distinct hot and cold splodges speckle the sky, betraying the slight variations in density and temperature. The significance of these variations is that they constitute the seeds of the large-scale structure of the universe. They were slowly amplified over time as the gravitational pull of the over-dense regions drew in more matter from their environs. Eventually the denser blobs turned into clusters of galaxies.

From the point of view of life, the primordial irregularities revealed by WMAP are crucial. Had the variations been larger, the dense regions would have imploded to form gigantic black holes rather than galaxies. Conversely, weaker variations would have been inadequate to trigger the formation of galaxies. Either way, our very existence depends on the primordial density perturbations being about right. The origin of these perturbations is not understood, but a favourite theory is that they are the relics of quantum fluctuations during the inflationary era, forged in the first trillion-trillion-trillionth of a second after the cosmic origin, writ large by inflation and frozen in the sky. Why the amplitude is so bio-friendly is a matter of conjecture to which I shall return shortly.

It is not just the cosmic initial conditions that are surprisingly bio-friendly. The very laws of physics seem to be cunningly contrived for life – 'a put-up job', to use the words of Fred Hoyle. Hoyle was much struck by the fact that the formation of carbon inside stars is a very touch-and-go affair. Carbon is made in a nuclear process that involves three helium nuclei coming together more or less simultaneously and fusing to make a single carbon nucleus. Without this reaction, the route to carbon, and elements beyond such as oxygen, would be blocked. On the face of it, the reaction rate for triple-helium fusion seems hopelessly small. Fred conjectured that there must be a special amplification factor at work, and sure enough there is. On rare occasions, a nuclear reaction can be greatly boosted by an effect known as resonance. Resonance is familiar in the everyday world, an oft-quoted example being the way in which an opera singer's voice can shatter a wine glass. It happens because the wine glass is highly sensitized to waves of a certain frequency, enabling it to absorb lots of energy at that frequency. In the same way, quantum mechanics tells us that nuclear particles have a wave aspect, and if the frequencies of the wave and the absorbing nucleus can be matched, energy can be transferred much more efficiently. It turns out that there is a nuclear resonance at just the right energy for three helium nuclei, plus some thermal energy, to make carbon with high efficiency. Hoyle realized that this happy coincidence, without which life may never have existed, depends in turn on the strength of the force that binds nuclei together. Had the nuclear force been just a few per cent stronger or weaker, carbon would not have formed in abundance in the universe.

Since Hoyle's observations in the 1950s, many additional examples of the 'fine tuning' of the laws of physics in relation to life have been uncovered. For example, the dissemination of carbon in supernova explosions depends rather finely on the strength of another nuclear force, known as the weak force. If the weak force were stronger, the all-important pulse of neutrinos, responsible for blasting the outer layers of the star into space, would be trapped inside the imploding stellar core. On the other hand, had the weak force been weaker, it would not have gained sufficient purchase on the outer layers of the star to propel them into the interstellar void.

From examples like these scientists have become generally convinced that the existence of life – at least, life as we know it – depends rather sensitively on both the cosmic initial conditions and the precise form of the laws of physics. One way to envisage this is to imagine playing god and attempting to design a universe with all the basic features we observe. Suppose this god could twiddle a knob and make the strong force stronger, or twiddle another knob and make all electrons heavier, and so on. A simple analysis reveals that more than very minor tinkering with the knobs would probably fatally compromise the universe's ability to generate life, and hence observers like us who can look back on cosmic history and ponder on the significance of it all.

Disagreement sets in, however, concerning the implications of this pervasive and suspicious bio-friendliness. Is it just a lucky fluke that the laws of physics and cosmic initial conditions came out right for life? Or might it hint at some element of pre-conceived design – divine providence even? Could it be that the laws of physics could not be otherwise, that their form is unique, fixed by logical and mathematical self-consistency? There is no consensus on the answers to these questions, but an increasingly fashionable view is that the explanation for cosmic bio-friendliness lies with a type of lottery effect. The basic idea is that what we have been all along calling the universe is nothing of the sort. Rather, it is but an infinitesimal component in a much vaster system dubbed 'the multiverse'.

If we could take a god's-eye-view of the multiverse, our universe would be but a single 'bubble' of space amid limitless other bubbles, and the big bang in which our cosmos originated would be accompanied by other bangs in other parts, perhaps forming an unending sequence. The multiverse is not just an idle speculation, but the natural outcome of attempts to understand the origin of the universe as a physical process. Clearly, if the big bang is a consequence of physical law, then it would be able to happen more than once, by definition. A truly unique process cannot be described as law-like; rather, it would be either a fluke or a miracle. One possible physical process for generating universes via big bangs is inflation. In the fashionable scenario due to Andrei Linde of Stanford University, the god's-eye-view would reveal an eternally inflating superstructure in which space stretches phenomenally fast –

doubling in volume every ten-billionth of a trillion-trillionth of a second on average. But amid this frenetic swelling, bubbles of space would separate out and follow a more sedate expansionary history. In this theory, our universe is one – but merely one – such cosmic region.

Linde's eternal inflation would not solve the riddle of cosmic bio-friendliness if all the bubble universes were more or less the same. Things get interesting, however, if different bubbles are characterized by different physics. How can this be? Surely physics is physics – it is universal? Perhaps not. In recent years, physicists have come to suspect that many previously sacrosanct 'laws' might not be truly fundamental at all, but more akin to frozen accidents. As the universe cooled from the searing heat of the big bang, so some familiar patterns of the microworld, such as the various particle masses and the strengths of forces, might have assumed arbitrary values, in much the same way that frost forms arbitrary patterns on the window when water freezes. If so, then different bubbles might have cooled with different values of these parameters, or even with totally different laws altogether. In that case what we have until now been regarding as universal laws of physics might instead be merely local by-laws, valid in our cosmic region but not beyond. (Our cosmic region is taken to extend at least several billion light years; there is little evidence that physics is significantly different in regions of the universe even at the furthest reach of our instruments.)

The idea of laws varying from one bubble universe to another has been greatly boosted by recent attempts to unify the various particles and forces of nature. The fashionable string/M theory, for example, predicts that there are at least 10^{500} different low-energy states, each producing different physics at low temperatures as the universe cools. So if eternal inflation pumps out an unending supply of universes from random initial states, then this stupendous variety of cosmic contenders will all eventually be sampled. The vast majority of 'bubble universes' will not enjoy low-energy physics consistent with life. But here and there, just by chance, there will be a universe with bio-friendly properties. Obviously the bio-hostile universes will go unobserved, so it is no surprise that we find ourselves living in a universe well suited to our existence – it could hardly have been otherwise.

This is not the place for me to weigh the arguments for and against the multiverse explanation of bio-friendliness, except to remark that, whatever its success or credibility, the multiverse theory still requires one to accept the existence of some form of physical law – albeit at a deeper level than the laws of our textbooks. At the very least, there has to be a law-like universe-generating mechanism. Also required are a host of other assumptions, such as the efficacy of mathematics in describing nature, the existence of discrete spacetime dimensions, and the unassailability of quantum mechanics, at least in the fashionable string/inflation version of the multiverse. That is, the multiverse does not exhaust all logical possibilities – far from it. An ultimate explanation for physical existence has to address the fact that something must separate what actually exists (even if what exists is far more than meets the eye) from what is possible but not in fact instantiated. That is, there must exist a rule that divides the real from the merely possible or, to use Stephen Hawking's words, to 'breathe fire into the equations' describing the world (or the multiverse). But one can then ask where the rule came from, and why that rule applies rather than some other. So whilst some form of multiverse seems both desirable and inevitable, given the evidence of fundamental physics and cosmology, it would be wrong to claim that the multiverse provides a complete explanation for why the universe permits life.

Whether cosmic bio-friendliness is regarded as a simple selection effect from a random sample, or as something more subtle and meaningful, the big bang that initiated our own cosmic history clearly has a creative aspect in both the cosmological and biological senses. However, the proliferation of big bangs postulated by the multiverse theory is alarming. Could another big bang happen in our cosmic backyard, destroying not just life, but the universe as we know it? A variety of cosmic catastrophes has been studied – explosions, implosions and abrupt transformations of universe-destroying proportions. Thankfully, no such process seems inevitable, or even likely, given the present state of our knowledge.

Are we alone? Violent aliens

In this final section I ask just how bio-friendly the universe might be. Life is undeniably a fragile phenomenon. Human life is especially precarious, and it is not hard to envisage ways in which humanity might be snuffed out by natural or made-made disasters. The titanic forces at work in the cosmos dwarf even the most ambitious human technological enterprise, and our much-vaunted intelligence would count for nothing in the face of even an asteroid impact. Placing humanity in a cosmic context, we seem overwhelmingly puny and vulnerable. Earth's entire biosphere amounts to less than one-billionth of its mass; our planet is a small body orbiting a dwarf star among hundreds of billions in our galaxy alone.

Yet masses and forces are not the only measure of worth. Humans possess one property that has so far been found unique in the universe: we have a measure of understanding of the world and a degree of control over events. How much does a brilliant thought weigh? I took the trouble to calculate it: the answer is about 10^{-80} of the mass of the observed universe. But one brilliant idea can eventually transform the world about us. The universe is nearly 14 billion years old. Human beings have existed in their present form for only 100,000 years or so. In that time – and mostly in the last few hundred years – we have changed the face of our planet, and even set foot on another astronomical body. The sun has at least another billion years before it will make Earth an uncomfortable planet, and there will be other star systems forming after that. If humanity survives the threats of the next few centuries, we can look forward to a vast longevity.

A popular scenario is that human beings overcome their innate destructiveness and either they, or their robotic or genetically engineered successors, spread out across the galaxy, colonizing suitable planets. It may take only a few tens of millions of years for our descendants to establish themselves galaxy-wide, but this is nothing compared to astronomical time-scales. If present astronomical evidence is to be believed, the universe will go on expanding for trillions of years, allowing time for intelligent life to gain control over arbitrarily large regions of space. In time, mind could rival brute cosmic forces in reshaping the cosmos.

All this supposes, however, that the galaxy and beyond are ours for the taking. But are we alone in the cosmos? Did the same bio-friendliness that spawned life on Earth work its magic on other planets too? Is life, even intelligent life, widespread in the universe? At this time, we have absolutely no scientific evidence one way or the other for any form of life beyond Earth, let alone intelligent life. It may be that we are totally alone in the universe, and our planet is a freak. Equally, our galaxy and beyond might be teeming with life, and any attempt by humans to spread across the galaxy could bring us into conflict with indigenous intelligent life. What then? Star Wars?

Clashes between humans and aliens are an enduring science fiction theme. It would be a depressing prospect if our exquisitely bio-friendly universe spawned irredeemably aggressive intelligent life that sullied the majesty of the cosmos with tedious conflict. But such is the legacy of Darwinian evolution: survival of the fittest is a fundamental principle of nature. So could it be that the universe is violent both astronomically and biologically? Do star wars rage all around us as we innocently take our first faltering footsteps beyond planet Earth? Fifty years ago the physicist Enrico Fermi asked the question, Where are they? Fermi wondered what was to stop aggressive alien civilizations from spreading across the galaxy and conquering Earth. The fact that this had not happened, Fermi took as evidence that there are no such aliens; that we are alone.

Fermi may be right. But there is another interpretation. Aggressive colonization and exploitation may be a primitive urge restricted to newly evolved intelligent life. Those species that survive this dangerous stage may progress, perhaps by the use of genetic manipulation and machine intelligence, to a more benign phase in which the sanctity of life is taken seriously, and for whom interplanetary conflict may be anathema. This rosy view of cosmic biology has no scientific basis, but it is my profound hope and belief. To be sure, life inhabits a violent universe, and life itself can be mindlessly violent. But life can also be creative and nurturing. I can think of no more fitting a conclusion than to cite Charles Darwin, after whom this college and these lectures are named, who expressed it thus: 'The love for all living creatures is the most noble attribute of man'.

FURTHER READING

P. Davies, *The Origin of Life*. London: Penguin, 2003.

A. Linde, *Particle Physics and Inflationary Cosmology*. Amsterdam: Harwood Academic Publishers, 1990.

L. Marschall, *The Supernova Story*. Princeton: Princeton University Press, 1994.

M. Rees, *Before the Beginning: Our Universe and Others*. London: Simon & Schuster, 1997.

D. Steel, *Rogue Asteroids and Doomsday Comets*. New York: Wiley, 1997.

Notes on the contributors

Ms **Kate Adie** was the BBC's chief news correspondent from 1989 to 2002, reporting from several of the major conflict zones of recent years, including Tiananmen Square, the Gulf War, the conflicts in the former Yugoslavia, Albania, Rwanda, China and Sierra Leone. Her honours include three Royal Television Society awards, the BAFTA Richard Dimbleby Award, and the Broadcasting Press Guild's Award for Outstanding Contribution to Broadcasting. Her books include: *The Kindness of Strangers*, *Corsets to Camouflage: Women and War*, *From Our Own Correspondent* and *Nobody's Child: Who Are You When You Don't Know Your Past?*

Professor **Lisa Anderson** is Dean of International and Public Affairs at New York's Columbia University, and one of America's foremost scholars of the Middle East and North Africa, Her books include *The State and Social Transformation in Tunisia and Libya, 1830–1980*, *The Origins of Arab Nationalism* and *Transitions to Democracy*. She has testified before the Foreign Relations Committees of both the House and the Senate, and published commentary in the *New York Times*, the *Washington Post* and the *Los Angeles Times*.

Professor **Simon Baron-Cohen** is Director of Cambridge University's Autism Research Centre. He was awarded the British Psychological Society's Spearman Medal, and the American Psychological Association's McAndless Award, for outstanding contributions to research. His books include *Mindblindness*, *Autism: The Facts*, *Teaching Children with Autism to Mindread* and *The Essential Difference: Men, Women and the Extreme Male Brain*. He is also author of a DVD-ROM entitled *Mind Reading: The Interactive Guide to Human Emotions.*

Professor **William Brown** is a leading authority on relations of employment, minimum wage, trade unions and collective bargaining. He serves as member of the ACAS Panel of Arbitrators and on the ACAS Council, as well as the Panel of Arbitrators, and the Low Pay Commission. He chairs the National Fire Brigades National Joint Council Disputes Committee, and the Advisory Committee of the TUC's Partnership Institute. His publications include: *The Individualisation of Employment Contracts in Britain and Employment Relations in Britain: 25 years of the Advisory, Conciliation and Arbitration Service.*

Professor **Barry Cunliffe** is a leading scholar of European prehistory, who has conducted and published major excavations in Britain, France and Spain. He has served on many public heritage bodies including English Heritage, the Museum of London and the Mary Rose Trust. He has worked extensively on radio and television. In addition to his numerous academic monographs, his popular books and reference texts include *Rome and the Barbarians, Rome and Her Empire, The Celtic World, Origins, Greeks, Romans and Barbarians: Spheres of Interaction, The Oxford Illustrated Prehistory of Europe* and *The Ancient Celts.*

Professor **Paul Davies** is an acclaimed physicist, writer and broadcaster at the Australian Centre for Astrobiology at Macquarie University, Sydney. His research interests are in the fields of cosmology, quantum field theory and astrobiology. He has received an Advance Australia Award and two Eureka Prizes, the Kelvin Medal, the Royal Society Faraday Prize, and the Templeton Prize in 1995. His books include: *The Mind of God, Other Worlds, God and the New Physics, The Edge of Infinity, The Cosmic Blueprint, Are We Alone? The Fifth Miracle, The Last Three Minutes, About Time* and *How to Build a Time Machine.*

Professor **Andrew Fabian** is Royal Society Professor at the Institute of Astronomy, and the Vice Master of Darwin College, Cambridge. He has broad interests in high-energy astrophysics, and has worked on data from most orbiting X-ray satellites. Currently his main areas of research concern clusters of galaxies and accreting black holes. In addition to his scientific publications, he is a founding convenor of the Darwin lecture series, and editor of two previous series; *Evolution* and *Origins*.

Professor **David Haig** is an evolutionary geneticist/theorist at Harvard University who has pioneered research into conflict and conflict resolution within the genome and associated relationships between parents and offspring. Starting from an interest in the biology of birds and plants, his research has moved increasingly towards the human sphere, and some unexpected dynamics of sex, gender and parenthood. He is author of *Genomic Imprinting and Kinship*.

Professor **Martin Jones** leads a bio-archaeology group at Cambridge University's Department of Archaeology investigating the origins and development of the human quest for food, drawing on a range of techniques from biology to molecular science. Currently his main areas of research involve the tracking of early agricultural expansion through crop genetic markers, and the Palaeolithic origins of the modern human meal. His books include: *England Before Domesday*, *The Molecule Hunt: Archaeology and the Search for Ancient DNA* and *Traces of Ancestry*.

Professor **Richard Wrangham** of Harvard University's Museum of Comparative Zoology is one of the pioneers of close observation of primates in their natural habitat. He has spent three decades studying chimpanzee cultures in the wild and comparing chimp cultures to human ones. In the early 1970s, he was among the first primatologists to note the tendency of chimpanzees to conduct so-called lethal raids against neighbouring groups. He is fellow of the American Academy of Arts and Sciences, and his awards include the Rivers Medal and the Baron-von-Swaine Award. His books include *Chimpanzee Cultures* and *Demonic Males: Apes, and the Origins Of Human Violence*.

Index

Page numbers in *italics* indicate figures or tables.